INTENTIONAL HUMAN DOSING STUDIES FOR EPA REGULATORY PURPOSES

SCIENTIFIC AND ETHICAL ISSUES

Committee on the Use of Third Party Toxicity Research with Human Research Participants

Science, Technology, and Law Program

Policy and Global Affairs Division

NATIONAL RESEARCH COUNCIL
OF THE NATIONAL ACADEMIES

THE NATIONAL ACADEMIES PRESS
Washington, D.C.
www.nap.edu

THE NATIONAL ACADEMIES PRESS 500 Fifth Street, N.W. Washington, DC 20001

NOTICE: The project that is the subject of this report was approved by the Governing Board of the National Research Council, whose members are drawn from the councils of the National Academy of Sciences, the National Academy of Engineering, and the Institute of Medicine. The members of the committee responsible for the report were chosen for their special competences and with regard for appropriate balance.

This study was supported by Contract No. 68-W-02-071, between the National Academies and the U.S. Environmental Protection Agency. The views presented in this report are those of the National Research Council Committee on the Use of Third Party Toxicity Research with Human Research Participants and are not necessarily those of the funding agencies.

International Standard Book Number 0-309-09172-1 (Book)
International Standard Book Number 0-309-53098-9 (PDF)

Additional copies of this report are available from the National Academies Press, 500 Fifth Street, N.W., Lockbox 285, Washington, DC 20055; (800) 624-6242 or (202) 334-3313 (in the Washington metropolitan area); Internet, http://www.nap.edu.

Printed in the United States of America.

THE NATIONAL ACADEMIES

Advisers to the Nation on Science, Engineering, and Medicine

The **National Academy of Sciences** is a private, nonprofit, self-perpetuating society of distinguished scholars engaged in scientific and engineering research, dedicated to the furtherance of science and technology and to their use for the general welfare. Upon the authority of the charter granted to it by the Congress in 1863, the Academy has a mandate that requires it to advise the federal government on scientific and technical matters. Dr. Bruce M. Alberts is president of the National Academy of Sciences.

The **National Academy of Engineering** was established in 1964, under the charter of the National Academy of Sciences, as a parallel organization of outstanding engineers. It is autonomous in its administration and in the selection of its members, sharing with the National Academy of Sciences the responsibility for advising the federal government. The National Academy of Engineering also sponsors engineering programs aimed at meeting national needs, encourages education and research, and recognizes the superior achievements of engineers. Dr. Wm. A. Wulf is president of the National Academy of Engineering.

The **Institute of Medicine** was established in 1970 by the National Academy of Sciences to secure the services of eminent members of appropriate professions in the examination of policy matters pertaining to the health of the public. The Institute acts under the responsibility given to the National Academy of Sciences by its congressional charter to be an adviser to the federal government and, upon its own initiative, to identify issues of medical care, research, and education. Dr. Harvey V. Fineberg is president of the Institute of Medicine.

The **National Research Council** was organized by the National Academy of Sciences in 1916 to associate the broad community of science and technology with the Academy's purposes of furthering knowledge and advising the federal government. Functioning in accordance with general policies determined by the Academy, the Council has become the principal operating agency of both the National Academy of Sciences and the National Academy of Engineering in providing services to the government, the public, and the scientific and engineering communities. The Council is administered jointly by both Academies and the Institute of Medicine. Dr. Bruce M. Alberts and Dr. Wm. A. Wulf are chair and vice chair, respectively, of the National Research Council.

www.national-academies.org

COMMITTEE ON THE USE OF THIRD PARTY TOXICITY RESEARCH WITH HUMAN RESEARCH PARTICIPANTS

JAMES F. CHILDRESS, Cochair (IOM), John Allen Hollingsworth Professor of Ethics, Professor of Medical Education, and Director, Institute of Practical Ethics and Public Life, University of Virginia, Charlottesville, Virginia

MICHAEL R. TAYLOR, Cochair, Senior Fellow and Director, Risk, Resource, and Environmental Management Division, Resources for the Future, Washington, D.C.

JAMES V. BRUCKNER, Professor of Pharmacology and Toxicology, Department of Pharmaceutical and Biomedical Sciences, College of Pharmacy, University of Georgia, Athens, Georgia

ALICIA L. CARRIQUIRY, Associate Provost and Professor of Statistics, Office of the Provost, Iowa State University, Ames, Iowa

ELLEN WRIGHT CLAYTON, Rosalind E. Franklin Professor of Genetics and Health Policy; Director of Genetics and Health Policy Center; Senior Fellow, Institute for Public Policy Studies; Professor of Pediatrics; Professor of Law, Vanderbilt University, Nashville, Tennessee

JOHN DOULL, Professor Emeritus, Department of Pharmacology, Toxicology, and Therapeutics, University of Kansas Medical Center, Kansas City, Kansas

HENRY T. GREELY, C. Wendell and Edith M. Carlsmith Professor of Law; Professor, by courtesy, of Genetics; Co-director, Program in Genomics, Ethics, and Society and Program in Law, Science, and Technology, Stanford University Law School, Stanford, California

SIOBAN D. HARLOW, Associate Professor; Associate Director, International Institute; Department of Epidemiology, School of Public Health, University of Michigan, Ann Arbor, Michigan

LESTER B. LAVE (IOM), University Professor, the Harry B. and James H. Higgins Professor of Economics; Professor, Engineering and Public Policy; Director, Green Design Initiative; Co-Director, University Electricity Industry Center, Graduate School of Industrial Administration, Carnegie Mellon University, Pittsburgh, Pennsylvania

BERNARD LO (IOM), Professor of Medicine and Director, Program in Medical Ethics, University of California, San Francisco, California

THOMAS A. LOUIS, Professor, Department of Biostatistics, Johns Hopkins Bloomberg School of Public Health, Baltimore, Maryland

JOSEPH V. RODRICKS, Founding Principal, ENVIRON International Corporation, Arlington, Virginia

CHRISTOPHER H. SCHROEDER, Charles S. Murphy Professor of Law and Public Policy Studies and Director, Program in Public Law, Duke University Law School, Durham, North Carolina

ROBERT TEMPLE, Director, Office of Medical Policy, Center for Drug Evaluation and Research; Acting Director, Office of Drug Evaluation, U.S. Food and Drug Administration, Rockville, Maryland

Liaison to Science, Technology, and Law Program

DAVID KORN (IOM), Senior Vice President for Biomedical and Health Sciences Research, Association of American Medical Colleges, Washington, D.C.

Staff

ANNE-MARIE MAZZA, Study Director
MICHELLE C. CATLIN, Senior Program Officer
KATHI E. HANNA, Consultant
STACEY E. SPEER, Senior Project Assistant
SARA DAVIDSON MADDOX, Editor
CAMILLE M. COLLETT, Research Associate (through March 2003)
SUSIE BACHTEL, Administrative Assistant (through August 2003)

Acknowledgments

The committee gratefully acknowledges the contributions of many individuals who provided invaluable information and data for the study:

Stanley H. Abramson, Arent Fox Kinter Plotkin & Kahn, PLLC; John Adgate, University of Minnesota; Stephen W. Brimijoin, Mayo Clinic; Angus Cameron, Roslin BioCentre; Arthur L. Caplan, University of Pennsylvania; Neil Carmichael, Bayer CropScience; Gail Charnley, HealthRisk Strategies; Ian Chart, AMVAC Chemical Corporation; Shelley Davis, Farm Worker Justice Fund; Michael L. Dourson, Technology Excellence for Risk Assessment; Monty Eberhart, Bayer CropScience; William H. Farland, U.S. Environmental Protection Agency; Daniel D. Federman, Harvard Medical School; Penelope A. Fenner-Crisp, International Life Sciences Institute; Lynn R. Goldman, Johns Hopkins Bloomberg School of Public Health; Suzanne T. Ildstad, University of Louisville; Stephen L. Johnson, U.S. Environmental Protection Agency; William Jordan, U.S. Environmental Protection Agency; Jeffrey P. Kahn, University of Minnesota; Nancy E. Kass, Johns Hopkins University Bloomberg School of Public Health; William Kelly, Center for Regulatory Effectiveness; Steven Lamm, Consultants in Epidemiology and Occupational Health, Inc.; Philip J. Landrigan, Mt. Sinai School of Medicine; Alan H. Lockwood, State University of New York; Judith A. MacGregor, Toxicology Consulting Services; Susan Makris, U.S. Environmental Protection Agency; Karen M. Martin, U.S. Environmental Protection Agency; Ray S. McAllister, CropLife America; Ernest E. McConnell, ToxPath, Inc.; Deidre L. Murphy, U.S. Environmental Protection Agency; Herbert L. Needleman, University of Pittsburgh Medical School; Erik D. Olson, Natural Resources Defense Council;

Jacqueline Patterson, Technology Excellence for Risk Assessment; Christopher J. Portier, National Institute of Environmental Health Sciences; Peter W. Preuss, U.S. Environmental Protection Agency; Jennifer Sass, Natural Resources Defense Council; Rita S. Schoeny, U.S. Environmental Protection Agency; Vera Hassner Sharav, Alliance for Human Research Protection; Bob Sielken, Sielken & Associates Consulting, Inc.; Christina B. Swartz, U.S. Environmental Protection Agency; Abraham J. Tobia, Bayer CropScience; Richard Wiles, Environmental Working Group; and Chris F. Wilkinson, C. Wilkinson, LLC.

This report has been reviewed in draft form by individuals chosen for their diverse perspectives and technical expertise, in accordance with procedures approved by the NRC's Report Review Committee. The purpose of this independent review is to provide candid and critical comments that will assist the institution in making its published report as sound as possible and to ensure that the report meets institutional standards for objectivity, evidence, and responsiveness to the study charge. The review comments and draft manuscript remain confidential to protect the integrity of the deliberative process.

We wish to thank the following individuals for their review of this report: William Cain, University of California, San Diego; Daniel Farber, University of Minnesota; Penelope Fenner-Crisp, International Life Sciences Institute; Lynn Goldman, Johns Hopkins University; Bernard Goldstein, University of Pittsburgh; Fernando Guerra, San Antonio Metropolitan Health District; Carol Henry, American Chemistry Council; David Hoel, Medical University of South Carolina; Jeffrey Kahn, University of Minnesota; Mary Faith Marshall, Kansas University Medical Center; Thomas McGarity, University of Texas; Gary Rischitelli, Oregon Health and Science University; and Jonathan Samet, Johns Hopkins University.

Although the reviewers listed above have provided many constructive comments and suggestions, they were not asked to endorse the conclusions or recommendations, nor did they see the final draft of the report before its release. The review of this report was overseen by R. Alta Charo, University of Wisconsin, and Donald Mattison, National Institutes of Health. Appointed by the National Research Council, they were responsible for making certain that an independent examination of this report was carried out in accordance with institutional procedures and that all review comments were carefully considered. Responsibility for the final content of this report rests entirely with the authoring committee and the institution.

Preface

One of the Environmental Protection Agency's (EPA's) most important and difficult tasks is regulation of the use of chemicals in order to protect human health and the environment. For several years it has struggled to determine whether in discharging this responsibility it should consider, accept, and rely on data from third-party studies that deliberately expose humans to toxicants. In accepting EPA's request for assistance in resolving this question, the National Academy of Sciences established an interdisciplinary committee under the auspices of the Science, Technology, and Law Program to prepare a report and make recommendations.

EPA's difficulty—and the committee's difficulty—in resolving this question arises from several conflicting interests and social values. For many the idea of research that intentionally exposes humans to toxicants is repugnant. Yet for others the potential of such studies to define more accurately human risk makes them worthwhile and acceptable. Recognizing the range of views, the committee proceeded to analyze and assess the arguments for and against intentional human dosing studies. Supporters of these studies see them as similar to Phase 1 drug trials, which expose participants to chemicals without the prospect of direct benefit to them and which can be ethically justified if they assiduously follow certain standards and procedures, such as those embedded in the Common Rule. However, critics contend that third-party studies conducted for EPA regulatory purposes, but not conducted or sponsored by EPA, usually fail to follow those standards and procedures. Defenders of human dosing studies hold that they can provide scientific data that are more relevant for regulatory purposes than data from animal studies, while critics charge that most of these studies lack scientific validity and, in any event, see little benefit in raising acceptable limits for toxicants, the usual purpose of such studies. These conflicting views represent a sample of the concerns, found among members of the committee as well as in society at large, that have animated the debate.

Like other groups that have addressed this subject, we note that such testing should be approached with the utmost caution and care. This committee consisted of members with expertise in ethics, law, pharmacology, toxicology, genetics, pediatrics, statistics/biostatistics, economics, epidemiology, risk assessment, and clinical trials. We met 6 times over 12 months in open and closed sessions and invited testimony from a number of individuals. In addition, we convened one public forum on January 8, 2003, to receive public input on the topics under consideration. During the course of the study, we received and reviewed voluminous studies voluntarily provided by a number of pesticide companies that had previously conducted intentional dosing studies and submitted their results to EPA for consideration. Some of these submissions were complete files on a particular chemical, while others were partial files. All of these materials were placed in the National Academies' public access file for this project. In addition, committee staff filed a Freedom of Information request with EPA for all information relevant to the intentional dosing studies that had been submitted to the Office of Pesticide Programs. Committee staff reviewed these studies and briefed the full committee on their findings.

In addressing EPA's questions, members of the committee read and listened carefully and thought rigorously and imaginatively about the recommendations that could be made and the rationale for those recommendations. As a result, every member of the committee changed his or her mind on some important topic in this report in the course of the extended and intense discussions and deliberations. As cochairs of this committee, we express our deep gratitude to committee members and staff, all of whom devoted enormous time and intellectual energy to the development of this report.

James F. Childress and Michael R. Taylor
Cochairs

Contents

APPENDIXES

List of Tables and Boxes

TABLES

BOXES

Acronyms

AChE	Acetylcholinesterase
DHEW	U.S. Department of Health, Education, and Welfare
DHHS	U.S. Department of Health and Human Services
DMC	Data Monitoring Committee
DSMB	Data and Safety Monitoring Board
EPA	U.S. Environmental Protection Agency
FDA	U.S. Food and Drug Administration
FFDCA	Federal Food, Drug, and Cosmetic Act
FIFRA	Federal Insecticide, Fungicide, and Rodenticide Act
FQPA	Food Quality and Protection Act
GCP	Good Clinical Practice
GI	Gastrointestinal
HED	Human Equivalent Dose
IND	Investigational New Drug
IOM	Institute of Medicine
IRB	Institutional Review Board
IRIS	Integrated Risk Information System

LOAEL	Lowest Observed Adverse Effect Level
LOEL	Lowest Observed Effect Level
MTD	Maximally Tolerated Doses
NAS	National Academy of Sciences
NBAC	National Bioethics Advisory Commission
NCQA	National Committee on Quality Assurance
NIH	National Institutes of Health
NOAEL	No Observed Adverse Effect Level
NOEL	No Observed Effect Level
NRC	National Research Council
OHRP	Office for Human Research Protections
OIG	Office of Inspector General
OP	Organophosphates
OPRR	Office for Protection from Research Risks
PBPK	Physiologically-Based Pharmacokinetic
PD	Pharmacodynamic
PHS	U.S. Public Health Service
PK	Pharmacokinetic
RAC	Recombinant DNA Advisory Committee
RfC	Reference Concentrations
RfD	Reference Dose
SAB	Science Advisory Board
SAP	Science Advisory Panel
TSCA	Toxic Substance Control Act
UF	Uncertainty Factor
VOC	Volatile Organic Chemicals

Executive Summary and Recommendations

EXECUTIVE SUMMARY

The regulation of chemicals to protect human health and the environment is one of the Environmental Protection Agency's (EPA's) most important and controversial tasks. Chemicals play a central role in our modern industrial society and are pervasive in the environment and food supply. All chemicals have the potential to harm human health, depending on the conditions under which people are exposed. This makes it critically important from a public health perspective to understand the hazards and to control human exposures to chemicals so that risk of harm can be minimized or eliminated—the widely accepted purpose of chemical regulation. In practice, however, the regulation of the use of chemicals is controversial because it involves competing interests and values.

EPA administers a series of congressional enactments that establish basic standards and procedures for assessing and balancing the risks and benefits of chemicals through the regulatory process. Some of the most important issues with which EPA must grapple on a continuing basis involve the nature of the scientific evidence that will be acceptable and that will suffice as the basis for regulatory decision making.

EPA commissioned The National Academies to provide advice on the vexing question of whether and, if so, under what circumstances EPA should accept and consider intentional human dosing studies conducted by companies or other sources outside the agency (so-called third parties) to gather evidence relating to the risks of a chemical or the conditions under which exposure to it could be judged safe. EPA asked the committee to consider: (1) the conditions for which EPA should accept, consider, or rely on third-party, human toxicity studies (see Chapters 3-7); (2) under what circumstance(s), if any, the availability of human data should lead EPA to consider reducing or removing the customary 10-fold interspecies uncertainty factor (see Chapter 7); (3) the applicability of existing standards (e.g., the Common Rule, the *Declaration of Helsinki*) for evaluating the design and the conduct of this type of research (see Chapters 2 and 5); (4) whether and if so how the requirements of the Common Rule should be extended to the conduct of third-party human studies intended for submission to EPA in support of a regulatory decision (see Chapters 4-6); and (5) the extent to which, and how, the submitter of research with human subjects should be required to document or otherwise demonstrate compliance with appropriate standards for the protection of human research participants (see Chapters 3, 5, and 6).[1] The organization of this report has been a challenge because the issues and analysis are so intertwined. An effort has been made to provide a coherent narrative, but it has been necessary to make numerous cross-references among chapters.

The primary impetus for EPA's request was a series of events involving agricultural pesticides and EPA's implementation of the 1996 Food Quality and Protection Act (FQPA). This law modernized the safety standards applicable to pesticide residues in food, adding an extra measure of protection for children and placing strict deadlines on EPA's congressionally mandated program to ensure that all agricultural pesticides currently on the market satisfy the updated safety standards. The enactment and anticipated implementation of FQPA brought into question whether current uses of certain categories of long-used pesticides—the organophosphates (OPs) and carbamates—could be maintained under the new standards.

As a general rule, EPA sets safe levels of exposure to pesticide residue in food on the basis of extensive testing in animals to determine its toxic properties and to derive a Reference Dose (RfD). It then divides the highest dose at which the most sensitive indicator of human risk did *not* occur (the no observed adverse effect level or NOAEL) by two or more uncertainty factors to yield the relevant RfD. One uncertainty factor accounts for the possibility that the average human could be more sensitive to the chemical's effects than the animal model from which the NOAEL was identified (the interspecies factor). A second factor accounts for the possibility of variation among humans in their sensitivity to the chemical (the

[1]The complete charge to the committee is stated in Chapter 1, p. 40.

intraspecies factor).[2] EPA then makes its decision with regard to the FQPA mandate, which requires it to apply up to an additional 10-fold factor to take into account the potential for increased sensitivity for fetuses and children. The statute allows EPA to apply a factor other than 10 (i.e., lesser or greater) if reliable data are available to show that this different factor is protective of infants and children. The cumulative effect of this approach to determining safe levels of exposure to pesticides is a potential 1,000-fold margin of safety between the NOAEL in animals and allowable exposures in humans. It has long been EPA's practice to adjust the interspecies and intraspecies uncertainty factors if justified by scientific evidence showing that a different factor would provide a more scientifically sound or "accurate" extrapolation from the animal test results.

In response to FQPA, several pesticide manufacturers conducted and submitted to EPA intentional oral dosing studies involving humans for purposes of determining a NOAEL that might justify the reduction or elimination of the interspecies safety factor for certain pesticides in the widely used OP and carbamate classes. The submission of these studies has generated substantial controversy. Although it is not unusual or controversial for EPA to rely on human-derived data in its risk assessments, such data are typically derived from case reports, observational studies, or epidemiological studies that do not involve intentional dosing of humans.

In part, the pesticide studies involving humans are controversial because they were conducted by economically interested third parties, whose motivation was to justify reducing the interspecies uncertainty factor, thereby increasing the acceptable or safe human exposure level and possibly permitting the continuation of certain pesticide uses that might otherwise have been precluded under FQPA's new safety standards. Some scientists and environmental and other public interest groups challenged the ethical and scientific validity of the studies, contending among other things that people should not be put at risk for the purpose of reducing

[2]Application of additional uncertainty factors in deriving the RfD may be necessary (1) to account for the lack of chronic data if deriving a traditional, chronic RfD (i.e., the subchronic-to-chronic factor), (2) to extrapolate from a LOAEL (lowest observed adverse effect level) to an estimated NOAEL, if no appropriate NOAEL can be identified in the toxicity database (the LOAEL-to-NOAEL factor), or (3) to account for the absence of key data in the toxicity database for a given chemical (the database factor). The default values for the inter- and intraspecies uncertainty factors are 10; those for the other three generally range from 3 to 10. Of course, EPA has the discretion to modify any of these default uncertainty factors if justified by the available scientific evidence.

the stringency of regulatory standards. Pesticide manufacturers and some scientists argued that the human dosing studies were needed to ensure the scientific quality and accuracy of EPA's safety evaluations and that they had been or could be conducted ethically.

In response to this controversy, EPA declared in 1998 its intention not to use the pesticide studies until the ethical and scientific issues had been resolved and referred the matter initially to a Joint Subcommittee of EPA's Science Advisory Board (SAB) and the Federal Insecticide, Fungicide, and Rodenticide Act (FIFRA) Scientific Advisory Panel (SAP). The majority of the SAB/SAP Subcommittee concluded that there are circumstances in which such studies could be justified ethically and scientifically, subject to stringent conditions and oversight. (A minority report was filed by two members indicating that they could not envision a situation where these types of studies could be conducted.) After receiving the subcommittee's report, EPA decided to seek further review from the National Academy of Sciences (NAS) and to broaden the issue to encompass third-party human dosing studies related not only to pesticides but also to EPA's other chemical regulatory programs, including those addressing toxic air pollutants and drinking water contaminants.

EPA also asked the committee to consider the scientific basis that an otherwise ethically sound human study could rely on to alter the interspecies uncertainty factor. This important science policy issue lies behind much of the controversy surrounding pesticide studies.

Summary of the Committee's Response to EPA's Questions

The committee understands and respects both the intellectual difficulties and the social sensitivities involved in considering the issues surrounding human testing of chemicals. Like other groups that have addressed this subject, the committee noted that such testing should be approached with the utmost caution and care. Human studies involving pesticides, air pollutants or other toxicants, as compared to drugs or other therapeutic agents, are especially sensitive and controversial. To many, they are inherently repugnant and should never be allowed; to others, they may contribute significantly to science-based decision making.

Even though the tasks EPA assigned to this committee required that members consider difficult issues, the committee was not required to invent the basic standards that govern human research in the United States. These standards are already embodied in the Federal Policy for the Protection of Human Subjects (the Common Rule), which governs human research sponsored by EPA and many other government agencies, and in other authoritative statements of principle on the ethical conduct of human research. The committee's task was to consider how those standards should be applied in the particular case of intentional human dosing studies conducted by third parties for EPA regulatory purposes.

In keeping with these standards, the committee recommended that intentional dosing studies in humans be conducted and used for EPA regulatory purposes only if all the following conditions are met:

- The study is necessary and scientifically valid—that is, it addresses an important regulatory question that cannot be answered with animal studies or nondosing human studies and has been designed, conducted, and reported in a manner that ensures the study will be adequate scientifically to answer the question.
- The societal benefits of the study outweigh any anticipated risks to participants.
- Intentional human dosing studies that are to be used only to improve the accuracy of an RfD, and that otherwise provide no health or environmental benefit, can be justified only when there is reasonable certainty that participants will experience no adverse effects.
- All of the recognized ethical standards and procedures for protecting the interests of study participants are observed, including equitable selection and recruitment of participants, informed consent, and independent review of the scientific and ethical merits of the study by an Institutional Review Board (IRB) or its foreign equivalent.

The committee also recommended that EPA establish a high-level advisory board to conduct its own review of human dosing studies conducted for EPA regulatory purposes, whether sponsored by EPA or by a third party, both prior to and after the conduct of the study. The purpose of this review would be to ensure that the unique scientific and ethical issues associated with EPA-related studies, including whether the study is likely to be scientifically valid and otherwise beneficial for EPA regulatory purposes, have been thoroughly evaluated in advance by EPA and again prior to using the results of the study for regulatory purposes.

The committee carefully considered whether the studies on pesticides that, in part, gave rise to the request for this report provide societal benefits that should be considered in assessing their ethical acceptability. The committee concluded that in order to generate societal benefits such human dosing studies must: (1) be performed in a context in which there is a clearly defined regulatory objective and a critical, unanswered question or other compelling scientific need that cannot be satisfied with animal data or nondosing human studies and (2) be designed with the requisite statistical power and other design features required to meet that regulatory objective and scientific need. These are minimum threshold requirements that any human dosing study must meet.

Studies that satisfy this threshold test have the ability to improve the accuracy of EPA's regulatory decision making. The committee concluded that improving the accuracy of the science employed in regulatory decisions, whichever direction (i.e., lower or higher) it moves the RfD, constitutes a societal benefit that can justify the conduct of a human dosing study. If the intent is to raise the RfD, however, such a study is justified only if there is no identifiable risk to participants, (as in some pharmacokinetics [PK] studies that are expected, based on very low dose levels and extensive animal testing, not to cause any biological effect in study participants), or there is a reasonable certainty, grounded in the careful review of a sufficient body of scientific evidence, that participants will experience no harm (in the sense of impairment or pain), whether lasting or transitory.

Beyond the threshold benefit of improving scientific accuracy (and raising an RfD), human dosing studies can generate different kinds of societal benefits as well, such as benefits to human health or the environment, depending on the nature of the scientific question a study seeks to answer, the uses to which the study results may be put, and the consequences that may flow from those uses. In cases in which such additional benefits are present, they can be considered in determining the extent of the potential risk to which participants may justifiably be exposed, and such additional health or environmental benefits are required to justify the consideration of a human dosing study that is not in the "no identifiable risk" or "reasonable certainty of no harm" categories. Even when such additional benefits exist, a human dosing study that could be anticipated to cause lasting harm to study participants could not be ethically justified.

Finally, although the committee's charge was directed to third-party human dosing studies, the committee concludes that the ethical and scientific issues are fundamentally the same whether a human study is conducted by a third party or by EPA and that the same basic ethical framework should apply to both categories of studies.

Because of the complexity of the issues considered by the committee and the need to be specific about the proposals being made, the recommendations follow.

RECOMMENDATIONS

The committee makes 17 recommendations to strengthen oversight and provide guidance for the use of intentional human dosing studies at

EPA. These recommendations are directed to EPA, IRBs, and research sponsors/investigators.[3]

Establishing Scientific Acceptability

The scientific and ethical considerations of human participants' research are closely related. Research that deliberately exposes humans to toxicants must be both scientifically and ethically justified. Such a study could be scientifically valid but ethically unacceptable (e.g., because the investigator failed to get informed consent or exposed participants to too much risk); however, a study cannot be ethically acceptable if it is scientifically invalid. A sound research design is the first step in developing an ethically acceptable protocol. For these reasons, scientific and ethical considerations should be integrated in the review and evaluation of all human research studies.

Recommendation 3-1: Scientific Validity of Intentional Human Dosing Studies

EPA should issue guidelines for determining whether intentional human dosing studies have been:

a. justified, in advance of being conducted, as needed and as scientifically appropriate, in that they could contribute to addressing an important scientific or policy question that cannot be resolved on the basis of animal data or human observational data;
b. designed in accordance with current scientific standards and practices to (i) address the research question, (ii) include representative study populations for the endpoint in question, and (iii) meet requirements for adequate statistical power;
c. conducted in accordance with recognized good clinical practices, including appropriate monitoring for safety; and
d. reported comprehensively to EPA, including the full study protocol, all data produced in the study (including adverse events), and detailed analyses of the data.

Balancing Risks and Benefits

Even if scientifically valid, an intentional human dosing study is not ethically acceptable unless the benefits it provides to the participants or to

[3]The order of the recommendations in this executive summary does not match the order in which they appear in the full report. For clarity, in this summary, the recommendations are provided in an order more conducive to a shorter format with abbreviated discussion.

society outweigh any risks posed to participants. Risks will vary widely depending on the inherent properties of the chemical and the particular conditions of exposure. Careful assessment of risks to participants thus is a prerequisite for conducting a human dosing study. The committee identifies three principal types of human dosing studies conducted for EPA regulatory purposes, each involving different levels of risk based on the particular information sought: (1) those seeking PK information; (2) those studying effects on a biomarker but not adverse signs or symptoms; and (3) those studying adverse but reversible effects. None of the studies the committee encountered would be expected, based generally on extensive animal data and human experience, to cause any irreversible or serious adverse effects. Low-dose PK studies that are expected based on extensive animal testing not to cause any detectable biological response commonly pose no identifiable risk to participants. For the biomarker studies in the second category, there typically are sufficient data to conclude with reasonable certainty that no harm will occur to participants from the biomarker changes. Studies in the third category, because they cause adverse effects, pose an identifiable risk the seriousness of which could vary widely.

The potential benefits of an intentional human dosing study also can vary widely. Participants in human dosing studies conducted for EPA regulatory purposes are not likely to benefit personally from their participation, except to the extent they are paid for their participation. The committee concludes that financial remuneration is not a benefit that should be considered in balancing the risks and benefits of these toxicant studies, which means that the relevant benefits potentially associated with human dosing studies conducted for EPA regulatory purposes are societal. For example, a human dosing study on an air pollutant that provides essential data to establish or strengthen a health-protective standard confers on society a potentially significant health benefit. Likewise, a study that would make it possible for EPA to approve a pesticide intended to control a disease vector, such as mosquitoes or ticks, benefits society in a way that could properly be considered in balancing the risks and benefits of a study.

In light of the nature and purpose of the human dosing studies that prompted this report, one of the critical questions the committee addressed was whether an intentional human dosing study anticipated to improve the scientific accuracy of EPA's decisions—for example, by raising the RfD—but not to directly enhance health or environmental protection confers a societal benefit. The committee carefully considered the congressional judgments and intent underlying EPA's chemical regulatory programs, including the requirement that EPA use the best available scientific evidence in making its regulatory decisions.

The committee reviewed a number of intentional human dosing studies of the kind typically submitted to EPA or conducted by EPA for regulatory purposes, including several of the OP pesticide studies that prompted this report. Several studies reviewed by the committee measured cholinesterase inhibition, which has been widely studied in humans, as a biomarker of exposure and potential toxicity, rather than a toxic endpoint per se, and were conducted to support reduction of the interspecies uncertainty factor.

Recommendation 4-1: Value of Studies That Seek to Improve the Accuracy of EPA's Decisions But Do Not Provide a Public Health or Environmental Benefit

EPA should consider a human dosing study intended to reduce the interspecies uncertainty factor (for example, a study of a biomarker such as cholinesterase inhibition) as conferring a societal benefit only if it was designed and conducted in a manner that would improve the scientific accuracy of EPA's extrapolation from animal to human data. Because the anticipated benefit would not be as great as that conferred by studies intended to provide a public health or environmental benefit, the study could be justified ethically only if the participants' exposure to the pesticide could reliably be anticipated to pose no identifiable risk or present a reasonable certainty of no harm to study participants.

Recommendation 4-2: Value of Studies That Seek to Provide a Potential Public Health or Environmental Benefit

An IRB should be properly constituted to be able to consider whether a study has the potential of providing a clear health or environmental benefit to the community. Such studies could be acceptable even if they involved a somewhat higher level of risk than that posed by studies for which there is no identifiable risk or for which there is a reasonable certainty of no harm. No study is ethically justifiable if it is expected to cause lasting harm to study participants.

Ethical Considerations

Many ethical considerations remain after determining that a research protocol is scientifically valid and that its probable benefits outweigh its risks to research participants. These other ethical considerations include fair selection and recruitment of potential research participants, fair payment for their participation, the provision of voluntary informed consent, and the provision of compensation for research-related injuries.

Recommendation 5-1: Criteria for Scientific and Ethical Acceptability

Studies that do not meet the highest scientific and ethical standards should not be carried out or accepted by EPA as input to the regulatory decision-making process. Necessary conditions for scientifically and ethically acceptable intentional human dosing studies include:

a. prior animal studies and, if available, human observational studies;
b. a demonstrated need for the knowledge to be obtained from intentional human dosing studies;
c. justification and documentation of a research design and statistical analysis that are adequate to address an important scientific or policy question, including adequate power to detect appropriate effects;
d. an acceptable balance of risks and benefits and minimization of risks to participants;
e. equitable selection of participants;
f. free and informed consent of participants; and
g. review by an appropriately constituted IRB or its foreign equivalent.

• *Selection of Research Participants*

According to the Common Rule, IRBs should not approve a research protocol involving research participants unless "selection of subjects is equitable" (40 CFR 26.111(3)). The principle of justice directs attention to the *distribution* of benefits and risks—who will gain the benefits and who will bear the risks and other burdens of research—not just the overall risk-benefit ratio. Not only should the research participants be representative of the target population of interest, but the selection of participants should be inclusive in order to avoid exploitation of any particular social group. Particular concerns arise about the recruitment of persons from vulnerable populations, including persons who lack decision-making capacity and persons who may be vulnerable to coercion or undue influence.

Some potential participants may be at increased risk of harm from particular research protocols. In general, individuals who would face higher risks in the experiment should not be selected for participation. An exception might be warranted if their participation is necessary to answer a question of major importance in the regulatory process and perhaps one of special relevance to people with their condition. But, even then, additional protective measures would be required.

Children represent a special case. They are vulnerable because they lack decision-making capacity and are greatly influenced by adults and are often more susceptible to the adverse effects of toxicants. The Department of Health and Human Services (DHHS) has addressed the tension between the need for greater knowledge about children and the need to protect them from harm and exploitation in research. Subpart D (Additional DHHS Protections for Children Involved as Subjects in Research) greatly restricts the enrollment of children in research that involves greater than minimal risk without the prospect of direct medical or health benefit.

Recommendation 5-2: Participant Selection Criteria

IRBs reviewing intentional human dosing studies should ensure that the following conditions are met in selecting research participants:

a. Selection should be equitable.

b. Selection of persons from vulnerable populations must be convincingly justified in the protocol, which also must justify the measures to be taken to protect those participants.

c. Selection of individuals with conditions that put them at increased risk for adverse effects in such studies must be convincingly justified in the protocol, which also must justify the measures that investigators will use to decrease the risks to those participants to an acceptable level.

EPA should adopt Subpart D of the Regulations for the Protection of Human Research Subjects. At a minimum, EPA should adhere to Subpart D's requirements for research involving children.

• *Payment to Participants*

Another issue related to the principle of justice, as well as of respect for persons, involves remuneration for participation in research. Paying research participants is a common and long-standing practice in the United States. Ethically, the principles of justice, fairness, and gratitude support payment to those who bear the burdens of research on behalf of society. Nonetheless, there is little agreement in theory or in practice about what constitutes just or fair payment. Any remuneration will influence the decisions of some more than others, and the protocol must be careful to protect participants, even when they misrepresent their health state and symptoms in order to participate and receive payment. All parties involved in designing and evaluating a protocol should consider whether the proposed level of remuneration would constitute exploitation or offer undue inducement.

Recommendation 5-3: Payment for Participation

IRBs, all relevant review boards, investigators, and research sponsors should ensure that payments to participants in intentional human dosing studies are neither so high as to constitute undue inducement nor so low as to be attractive only to individuals who are socioeconomically disadvantaged. Proposed levels of and purposes for remuneration (e.g., time, inconvenience, and risk) should be scrutinized in light of the principles of justice and respect for persons.

Moreover, EPA, in conjunction with other federal agencies, should consider developing further guidance on remuneration for participation in intentional human dosing studies, including guidance regarding whether remuneration should reflect the level of risk as well as the time and inconvenience involved.

- *Informed Consent*

Voluntary, informed consent by research participants (or permission by their surrogate decision makers) is a principal requirement in the system of protections of research participants. The consent requirement expresses the principle of respect for persons, including their autonomous choices. The Common Rule stresses this requirement, as do other codes of research ethics, including the Nuremberg Code, the *Declaration of Helsinki*, and Food and Drug Administration's (FDA) Good Clinical Practice (GCP) guidelines. To ensure the voluntary, informed consent of participants in toxicant studies, the committee recommends the development of a list of best practices for the consent process. These practices should be used to stimulate investigators and IRBs to consider what consent procedures would be most appropriate for a particular study. They should not be regarded as inflexible requirements that must be applied in every case.

Recommendation 5-4: Best Practices in Informed Consent

EPA should develop and disseminate to relevant IRBs, investigators, and sponsors a list of best practices regarding informed consent in intentional human dosing studies. EPA should encourage

all sponsors and investigators to adopt these practices, and it should require their adoption in studies it sponsors or conducts.

- *Compensation for Research-Related Injuries*

Debate continues in the United States about whether compensation should be provided for research-related injuries. The Common Rule requires only that when research involves more than minimal risk, information should be disclosed about whether medical treatments and other compensation will be provided for research-related injuries. Many critics of the U.S. policy believe there should be more than disclosure of information about compensation, calling for provision of medical care for research-related injuries without cost to the injured participants and, in addition, for compensation for lost wages, disabilities, and death. These claims are based on the belief that research participants, whatever their motivations, accept risk on behalf of society. When research participants are injured, justice, fairness, and gratitude mandate, at a minimum, the provision of needed medical treatment without cost to the participant. Further study is needed regarding the provision of other compensation.

Recommendation 5-5: Compensation for Research-Related Injuries

At a minimum, sponsors of or institutions conducting intentional human dosing studies should ensure that participants receive needed medical care for injuries incurred in the study, without cost to the participants.

In addition, EPA should study whether broader compensation for research-related injuries should be required.

Creation of a Comprehensive EPA Human Studies Review Process

EPA is a signatory agency to the Common Rule, which requires, at a minimum, that human research protocols undergo review by an IRB and that participants provide voluntary informed consent. The Common Rule applies to human research sponsored by EPA as well as any research performed at an institution that has committed to have all research reviewed by an IRB as part of its assurance of compliance. Private sponsors of intentional human dosing studies submitted to EPA are not required by U.S. law to obtain IRB approval for studies, unless the studies are conducted at institutions that require IRB review of all research. However, it appears that all of the pesticide experiments reviewed by the committee were approved in advance by IRBs or their foreign equivalents. Even though the sponsors of those experiments acted responsibly in submitting their protocols for IRB review, this decision should not be left to the sponsors' discretion.

EPA itself has sponsored intentional human dosing studies involving exposure to toxicants. At least some of those experiments were approved by IRBs at the institutions that conducted the research. The committee was informed that EPA does not have an IRB, but instead has an Ethics Review Officer who typically ensures that all EPA-sponsored or conducted studies have been reviewed by an IRB. If all EPA-sponsored human research is conducted at nonfederal institutions and those institutions have appropriate IRBs operating in compliance with the Common Rule, the federal requirements might be satisfied. If EPA conducts human research in-house, it must continue to ensure that the research is reviewed by an appropriately constituted IRB.

Recommendation 6-1: IRB Review of All Studies

EPA should require that *all* human research conducted for regulatory purposes be approved in advance by an appropriately constituted IRB or an acceptable foreign equivalent. Research conducted by EPA scientists should be reviewed by an EPA-authorized IRB.

As noted above, IRBs remain a crucial part of the system of protection for participants in research. However, in special situations in which research poses complicated scientific and ethical issues, as in intentional human dosing studies, IRB review requires substantial supplementation. The committee concludes that another level of review is needed for intentional human dosing studies in order to add a supplementary layer of protection and to establish a body of knowledge and expertise with regard to these studies that can then be communicated to the public and the research community.

Recommendation 6-2: Human Studies Review Board

To ensure that intentional human dosing studies conducted for EPA regulatory purposes meet the highest scientific and ethical standards, EPA should establish a Human Studies Review Board to address in an integrated way the scientific and ethical issues raised by such studies. To the extent possible, this board should review in a timely manner the protocols and the justification for *all* intentional dosing studies intended for submission to EPA, as well as study results when completed. These reviews should be conducted regardless of the sponsor or site of performance, and EPA should communicate the results of the reviews to relevant parties.

The Human Studies Review Board should prospectively review the protocols and the justification for all studies, whether third party or EPA sponsored or conducted. While studies sponsored or conducted by EPA would be required to undergo review by the Human Studies Review Board in advance, private entities should be encouraged to voluntarily submit their protocols to the board before beginning a study. The committee notes that it would be optimal if this review of privately sponsored studies were mandatory, but because of legal and logistical concerns it recommended only that EPA consider making it mandatory. Any conclusions reached by the board should be advisory and not binding on the sponsoring companies or reviewing IRBs. The proposed board supplements but does not replace the IRB. Its principal function would be to help assure that EPA considers only intentional human dosing studies that meet the rigorous scientific and ethical standards specified in this report. Before human toxicant experiments are conducted, the board would provide advice to the sponsors proposing such research (including EPA) on how to meet these high standards. Furthermore, EPA's awareness of all studies would help ensure that when studies unexpectedly suggest that an environmental standard must be strengthened or that a safety factor must be increased, such studies would be included in the EPA regulatory or risk-assessment processes. After the experiments are completed and the results submitted to EPA, the board would advise EPA's relevant program offices on whether, and to what extent, the results should be considered. It would also, over time, collect and analyze information about these experiments that could enable it to suggest ways to improve such research or to assess whether EPA should continue to consider the results of these types of experiments.

The post-experiment review function of the board is distinct from the kind of review that EPA undertakes for the purpose of incorporating results from particular experiments into the regulatory process. It would not replace or modify the structures and procedures for the latter kind of review. Instead, it would offer nonbinding advice to the relevant EPA units about the scientific and ethical acceptability of the completed and submitted research.

Finally, the committee recommends a structure for review of these experiments that should be both rigorous and workable, but it recognizes its limits in foreseeing how well the structure might work over time and whether it will continue to be needed. Hence, timely periodic reviews will help ensure that the board plays the valuable role this committee envisions for it.

Recommendation 6-3: Review of the Human Studies Review Board

The proposed Human Studies Review Board, its functions, and its record should be assessed after 5 years by a body composed of EPA staff and external reviewers.

To review data submitted from intentional dosing studies for regulatory decision-making purposes (e.g., setting standards), EPA should provide sufficient and appropriate in-house expertise, at least at the level that exists for review of animal studies. The results of scientific review of data for regulatory purposes and its use in setting standards should be communicated to the board. It is the committee's view that the Human Studies Review Board is advisory only and is not a replacement for the scientific review EPA must perform in making regulatory decisions.

EPA's Use of Data from Studies of Cholinesterase Inhibition

The committee was asked to evaluate the use of data from intentional human dosing studies in EPA's risk-assessment process. Questions have arisen regarding the circumstances, if any, in which it would be appropriate to use such data, and the manner in which they should be used. The committee examined those questions within its task of considering whether and in what ways data from intentional dosing studies in humans could be appropriately incorporated into EPA's general framework for risk assessment. The committee was not asked to review the framework itself and does not offer an assessment of it in this report.

Recommendation 7-1: Review of Scientific Data

EPA's use of data from third-party intentional human dosing studies involving cholinesterase inhibition is advisable only if the agency undertakes a thorough review of the data (of the type typically undertaken for submitted animal studies and informed by external peer review) and finds that the studies substantially meet the scientific and ethical standards elucidated in this report. If the studies are found to be scientifically and ethically satisfactory, EPA should use the data to establish RfDs.

For those cholinesterase inhibitors that have been thoroughly investigated in high-quality animal studies (including studies of developmental neurotoxicity), and for which it is clear that cholinesterase inhibition is the most sensitive indicator of toxicity, data from intentional human dosing studies may be considered for use in risk assessment. It should be recognized that these circumstances—in which the most sensitive indicators of toxicity are the acute biological effects of chemicals and in which such effects are readily measurable in ethically acceptable human studies—are likely to be highly unusual. Indeed, at present the committee was not aware of other candidates for such studies. The committee's recommendations regarding the cholinesterase inhibition studies are thus not expected to suggest many other cases in which intentional dosing studies in humans to establish a NOAEL will be of value and therefore justifiable. The committee's recommendations regarding study justification (Recommendations 3-1, 4-1, and 5-1), in which proponents of intentional dosing studies in humans must document that the endpoints to be measured are the critical determinants of risk, represent a substantial hurdle.

Recommendation 7-2: Use of Existing Cholinesterase Inhibition Studies

The cholinesterase inhibition studies that already have been submitted to EPA, if determined to be scientifically valid and justified for EPA's regulatory purposes, may be considered for use in risk assessment and standard setting if they were not unethically conducted (see Recommendation 5-7).

As indicated in these recommendations, under stringent conditions data from intentional dosing studies in humans can be used within EPA's risk-assessment framework. Use of such data will eliminate the need for the uncertainty factor (UF_A) ordinarily used to extrapolate from animals to humans of average sensitivity. The safety factor called for under FQPA to protect children will not be affected by the use of data from intentional dosing studies in humans. Information directly relevant to children cannot be obtained from intentional dosing studies in human adults, and any such studies in children would be beyond ethical bounds.

Recommendation 7-3: Eliminating or Replacing the Interspecies Uncertainty Factor

In considering the use of data from the cholinesterase inhibition studies already submitted to EPA, the agency should clearly communicate to all stakeholders that information used to eliminate the interspecies uncertainty factor (UF_A) will have no influence on the use of other uncertainty factors or on the use of the safety factor protecting children as required by FQPA.

Several critical questions remain regarding the use of data from intentional dosing studies in humans. Studies that reveal no effects of any type at the doses used (so-called NOEL-only studies [no observed effect level]) may provide some data regarding safety, but they are inadequate for deriving RfDs or any other formal measure of human protection. Such data should be used only if there are no other data available and there is a compelling public health need to derive a tentative measure of public health protection because they provide no assurance that the study was capable of detecting the effect of interest. Moreover, the relationship between the presumed sensitivity of the study population and the presumed sensitivity of average humans is somewhat ambiguous and needs clarification. Thus, it is not completely clear that the people who participate in intentional dosing studies are always "individuals of average sensitivity" and that they are not, in fact, *less* sensitive than the "average." Uncertainties regarding these relationships may be dealt with by a requirement for study replication in a different setting, or by use of an uncertainty factor for intraspecies extrapolation (UF_H) that is somewhat greater than the usual default factor of 10.

Recommendation 7-4: Data from NOEL-Only Studies and the Sensitivity of Study Populations

EPA should reject data from NOEL-only studies for risk assessments if the NOEL is defined as the absence of any biological response, because such studies do not show levels that give rise to an effect (the LOEL [lowest observed effect level]). Such studies provide no assurance that they were adequate to detect the effect of interest. The agency also should consider whether the uncertainty factor used for intraspecies variability (UF_H) should be increased to deal with the possibility that study participants may be of less than average sensitivity. A request for study replication also should be considered as a way to address this last issue.

Use of Results from Ethically Problematic Studies

A final question concerns what role, if any, ethically problematic or unethical studies should play in EPA's regulatory decisions. The committee predicts that this question will rarely present itself after EPA formulates its new standards and procedures. However, when the question does arise in relation to such studies, it can raise difficult ethical issues. The committee concluded that, as a general rule, EPA should not use data from ethically problematic studies to inform its regulatory efforts.

In an extraordinary case, when data from ethically problematic studies appear to warrant a regulatory standard that would provide better protection for public health, the Human Studies Review Board may recommend that EPA convene a special, outside panel, which should reach its judgment by considering:

(1) whether the data are crucially important for protecting the public and

(2) whether the data cannot otherwise be obtained, with reasonable certainty within a reasonable period, without exposing additional research participants to the risk of harm.

Unless the panel can answer both questions affirmatively, it should recommend that EPA not consider or rely on the data in question. In order to strongly deter sponsors and researchers from conducting unethical studies, data from such studies should not be used to favor the sponsor's interests in loosening regulatory standards.

Recommendation 5-6: Studies Completed After Implementation of the New Standards

EPA should operate on the strong presumption that data obtained in studies conducted *after* implementation of the new rules[1] that do not meet the ethical standards described in this report will not be considered in its regulatory decisions. Under exceptional circumstances, studies that fail to meet these ethical standards may provide valid information to support a regulatory standard that would provide greater protection for public health. Under these circumstances, EPA should convene a special, outside panel, consisting of relevant experts and members of the public, to examine the cases for and against considering data from such studies.

Consideration of the use of data that were collected *before* the new standards are placed into effect raises particularly difficult issues. Although standards for the ethical conduct of research have been evolving, some are universal (e.g., the requirement not to intentionally harm research participants), and others have a long history. However, often it would be difficult, if not impossible, to obtain sufficient evidence to determine whether past studies, especially those in the distant past, met the ethical standards in place at that time.

Recommendation 5-7: Studies Completed Before Implementation of EPA's New Standards

EPA should accept scientifically valid studies conducted before its new rules[2] are implemented unless there is clear and convincing

---

[1]The committee uses the term "rules" informally to mean guidance, guidelines, policy, protocols, rules, or regulations.

[2]See footnote 1.

evidence that the conduct of those studies was fundamentally unethical (e.g., the studies were intended to seriously harm participants or failed to obtain informed consent) or that the conduct was deficient relative to then-prevailing ethical standards. Exceptional cases in which the Human Studies Review Board determines that unethically conducted studies may provide valid information to support a regulatory standard that would provide greater protection for public health should be presented to a special outside panel, described in Recommendation 5-6, for consideration.

This special panel should consider recommending the use of such data only with the additional requirement that the ethical concerns raised by the study are documented and made publicly available. The committee's recommendations apply to both third-party and government-sponsored studies, and they apply to the cholinesterase inhibition studies that were central to the considerations of this committee.

1

Introduction and Background

The regulation of the use of chemicals to protect human health and the environment is one of the most important and persistently controversial tasks assigned by Congress to the Environmental Protection Agency (EPA). EPA's regulation of chemicals is important because of the central role they play in our modern industrial society and their potential consequences for health. Over the last 50 years, tens of thousands of chemicals have been developed and introduced into the environment in the United States. In a typical year, more than 6 billion pounds of toxic chemicals are released by industrial facilities into the environment.[1] Another 1.3 billion pounds of pesticides are applied annually to agricultural fields, homes, gardens, schools, and other settings.[2] In addition, chemical air pollutants are emitted from sources such as fossil fuel combustion that are involved in providing energy for transportation, power plants, and other industrial processes. All chemicals have the potential to harm human health, depending on the conditions under which people are exposed, particularly dosage. Thus, it is critically important to understand the hazards of chemicals and to control human exposure to them.

The regulation of chemicals is persistently controversial because it involves competing values and interests. Although all chemicals can pose risks, most also provide benefits or result from beneficial activities. Agri-

[1]See EPA's 2001 Toxics Release Inventory at www.epa.gov/tri/tridata/tri01/press/executivesummarystandalone.pdf.

[2]Ibid.

cultural pesticides, for example, contribute to our abundant, safe, and relatively inexpensive food supply. Chemicals are used to produce a vast array of other consumer products from which people directly benefit. They also enter the environment as by-products of activities people value, such as the burning of gasoline and other fuels for transportation and the production of electricity for heating and lighting our homes. No one wants the chemical by-products that can pollute the air and water and contaminate food, but few are prepared to do without the relatively low-cost energy that makes life comfortable and convenient. The controversy that stems from the two-sided nature of the use of most chemicals often is intensified by disputes among commercial interests, environmental groups, and others with diverse points of view about the proper assessment, balancing, and allocation of the associated risks and benefits. EPA's chemical regulatory programs operate at the intersection of these competing interests and values. The agency administers a series of congressional enactments that establish basic standards and procedures for assessing and balancing the risks and benefits of the use of chemicals through the regulatory process. These statutes resolve some of the broader questions about how particular categories of chemicals are to be regulated, but they leave many controversial issues unresolved. Some of the most important issues with which EPA must grapple on a continuing basis involve the nature of the scientific evidence that will serve as the basis for regulatory decision making.

EPA commissioned this study to help address a particularly vexing issue concerning the acceptability and usefulness of scientific evidence. The issue is whether and under what circumstances EPA should accept from outside parties, and consider in its regulatory decision making, studies that involve the intentional dosing of research volunteers in order to gather evidence relating to the risks of using a chemical or the conditions under which exposure to it could be judged safe.

This issue is one that is multifaceted and difficult. It involves the complex interplay between important ethical concerns and scientific questions regarding the validity and usefulness of human studies for EPA's regulatory purposes. Like chemical regulation in general, the issue of human testing involves competing interests and values. And the issue has an emotional component. For many, the idea of testing pesticides and other industrial chemicals and chemical contaminants in humans is, on its face, repugnant. It is natural and appropriate to question whether and why such testing should be conducted and considered by EPA, especially when the study participants gain little or no direct benefit.[3]

[3]This report uses the terms "participant" and "subject" to refer to persons who participate in research. The term "subject" has been widely used for decades and appears in the federal

This committee understands and respects both the intellectual difficulty and the social sensitivities involved when considering the issues surrounding human testing of chemicals. Like most other groups that have addressed this subject, the committee recommends that any testing of chemicals in humans be approached with the utmost caution and care.

It is essential also to consider the specific circumstances in which human testing is proposed to be conducted for EPA regulatory purposes. In medicine, human testing of drugs, many of which are extremely toxic, is a well-established practice. It is regulated by the Food and Drug Administration (FDA) and governed by a set of principles and rules that are designed to respect and protect research participants, while making it possible for society to benefit from the knowledge that can be gained from such research. The initial reason for, and still the primary purpose of FDA's oversight of human drug studies, is the protection of research volunteers. This primary purpose governed this committee's consideration of the similar but distinct issues facing EPA. The committee can envision circumstances in which human testing of chemicals in the EPA context could satisfy ethical and scientific standards, but, as will be made clear in the following chapters of this report, such circumstances are highly circumscribed and require careful oversight.

The remainder of this chapter provides brief background material on EPA's regulation of chemicals, the potential role of intentional human dosing studies, the events in EPA's pesticide program that prompted this report, the prior EPA advisory panel review, EPA's policy regarding ethical oversight of human studies, EPA's charge to the committee, the National Academy of Sciences committee process, and the organization of this report.

---

regulations, but the term "participant" has become more common in recent years both internationally and in the United States (NBAC, 2001). Neither term is fully satisfactory as a label for those who are enrolled in research. The argument for using one term over the other hinges on the interpretation of the relationship between investigator and the individual enrolled in research and on the relevant values, such as respect for persons. The term "subject" seems to many to suggest that the individual is subjected to the investigator's action, is in an unequal relationship with the investigator, and is often passive. By contrast, the term "participant" appears to many to be too broad, because it could apply to the investigator as well as to the individual enrolled in research; furthermore, it does not adequately describe those who are enrolled in research by others, such as surrogate decision makers. NBAC argued for using the term "human participant" in order to be more respectful to the individuals who participate in research and to emphasize that individuals should be active, not passive, in the decision to enroll in research studies (NBAC, 2001, 32-33). This committee report uses both participant and subject but, in accord with contemporary usage, most often uses participant unless the context, such as federal regulations, dictates the use of subject.

EPA'S REGULATION OF CHEMICALS

EPA regulates chemicals under numerous legal statutes enacted over several decades for a range of purposes generally involving the protection of human health and the environment (Box 1.1 summarizes a number of the major statutes). This report focuses on the human health aspects of these laws. Each law addresses a different set of chemical regulatory problems in the manner deemed appropriate by Congress. Several require EPA to make decisions based on assessments of the health risks that the use of chemicals may pose, which necessitates the collection of the scientific data required to make the assessment.

Five of these statutory schemes are particularly relevant to the issue being addressed in this report and will be described briefly here. Three of them—those involving pesticides and toxic substances generally—impose requirements on parties outside EPA (hereafter called third parties) to conduct tests and compile and submit data to EPA concerning the potential risks of chemicals. The other two—involving toxic air pollutants and drinking water contaminants—place the burden on EPA to assemble data on health risks and possibly conduct health effect studies.

EPA regulates pesticides under two statutes. Under the Federal Insecticide, Fungicide, and Rodenticide Act (FIFRA), EPA decides whether and under what circumstances a pesticide can be applied to food crops or be used for other pest control purposes without resulting in unreasonable adverse effects on human health or the environment. Under the Federal Food, Drug, and Cosmetic Act (FFDCA), EPA evaluates the safety of pesticide residues in food and establishes tolerances (maximum legally permissible levels) for specific pesticides in specific foods based on its conclusion that consumption of foods containing residues at these levels will be safe, which Congress defines as "a reasonable certainty of no harm." Under both FIFRA and FFDCA, the sponsor of the pesticide bears the burden of proving to EPA that a pesticide satisfies the statutory standards for approval (called registration) and for the granting of a tolerance. To meet this burden, sponsors typically must conduct extensive testing of the pesticide in accordance with testing requirements and guidelines established by EPA.

The other EPA-administered law that can require parties outside EPA to conduct tests on chemicals and submit the results to EPA is the Toxic Substances Control Act (TSCA), which Congress enacted in 1976 to give EPA the ability to screen new chemicals and track the 75,000 industrial chemicals produced by or imported into the United States. EPA screens these chemicals and can require testing of those that may pose an environmental or human health hazard. EPA can ban the manufacture and import of those chemicals that pose an unreasonable risk. TSCA's

BOX 1.1
Legal Statutes for EPA Regulation of Chemicals

The Federal Insecticide, Fungicide, and Rodenticide Act (FIFRA) (7 U.S.C. §136 et seq.)[a] was enacted in 1964 and has been amended numerous times since, including significant amendments particularly relevant to this report in the form of the Food Quality Protection Act (FQPA) of 1996 (P.L. 104-170)[b] (discussed below). In its current form, FIFRA mandates that EPA regulate the use and sale of pesticides to protect human health and preserve the environment.

Under FIFRA, EPA registers pesticides for use in the United States and prescribes labeling and other regulatory requirements to prevent unreasonable adverse effects on health or the environment. Under the Federal Food, Drug, and Cosmetic Act (21 U.S.C. §371),[c] EPA establishes tolerances (maximum legally permissible levels) for pesticide residues in food.

In addition, FQPA requires EPA to reassess all pesticide and other ingredient tolerances and exemptions that were in effect as of August 3, 1996 (when FQPA was signed). This effort is designed to ensure that existing tolerances and exemptions meet the safety standard set by FQPA.

The Toxic Substances Control Act (15 U.S.C. §2601 et seq.),[d] which Congress enacted in 1976, gives EPA the authority to screen and track chemicals produced by or imported into the United States. This currently amounts to over 75,000 chemicals.

The Clean Air Act (42 U.S.C. §7401 et seq.)[e] provides the primary framework for protecting humans and the environment from the harmful effects of air pollution. It is the comprehensive federal law that regulates air emissions from area, stationary, and mobile sources. This law authorizes EPA to establish National Ambient Air Quality Standards for pollutants such as ozone and particulate matter to protect public health and the environment, as well as to establish other standards for hazardous air pollutants.[f]

Under the authority of the Safe Drinking Water Act of 1974 (42 U.S.C. §300f et seq.), EPA sets standards for approximately 90 contaminants in drinking water.

---

[a]Available at www.epa.gov/region5/defs/html/fifra.htm.
[b]Available at www.epa.gov/oppfead1/fqpa/backgrnd.htm.
[c]Available at www.fda.gov/opacom/laws/fdcact/fdctoc.htm.
[d]Available at www.epa.gov/region5/defs/html/tsca.htm.
[e]Available at www.epa.gov/air/oaq_caa.html.
[f]Available at www.epa.gov/air/criteria.html.

premanufacture notification program is the only mechanism available to EPA other than the pesticide programs for requiring third parties to conduct and submit studies on the risks that may be posed by the use of chemicals as a condition to entering the marketplace.

Although the chemicals covered by TSCA and the pesticide laws have clearly identifiable commercial sponsors that can be required to conduct studies and submit health risk data to EPA, this is not always the case with respect to the chemicals covered by the toxic air pollutant and drinking water contaminant laws.

Under the Clean Air Act, EPA regulates and seeks to minimize the harmful health and environmental effects of air pollution. A key component of the Clean Air Act is a requirement that EPA significantly reduce daily, so-called routine emissions of the most potent air pollutants: those that are known or suspected to cause serious health problems such as cancer or birth defects. The Clean Air Act refers to these pollutants as "hazardous air pollutants," but they also are commonly known as toxic air pollutants or simply as air toxics. As amended in 1990, the Clean Air Act requires EPA to use a "technology-based" and "performance-based" approach to significantly reduce emissions of air toxics from major sources of air pollution, followed by a "risk-based" approach to address any remaining, or residual, risks. Eight years after each technology-based standard has been adopted, EPA must assess the remaining health risks and, if necessary, adopt additional standards that address any significant remaining risk. In these cases, EPA can require industry to collect data and conduct tests, but the burden is on EPA to assess risks and set standards. The Clean Air Act also contains provisions regarding the testing of new fuel additives.

Under the Safe Drinking Water Act, as amended in 1996, EPA sets standards to restrict the presence of contaminants in drinking water to a level that "maximizes health risk reduction benefits at a cost that is justified by the benefits." The process for setting these standards includes a risk-assessment step to determine the maximum level of a contaminant in drinking water at which no known or anticipated adverse effect on the health of persons would occur and that would allow an adequate margin of safety. These health-based levels are goals; the enforceable limits are set with the goals as the starting point, but they also include consideration of the costs required to achieve a given reduction in health risk.

THE POTENTIAL ROLE OF HUMAN STUDIES

In implementing all of these chemical regulatory statutes, EPA follows standard approaches to risk assessment of chemicals. This includes the standard model of risk assessment described by the National Research

Council in 1983 (NRC, 1983; see Box 1.2 for a summary of the model as applied to pesticides), and the standard approach to evaluating the safety of chemicals in food that has evolved over the past 40 plus years in the food safety programs of both FDA and EPA. In both cases, for purposes of initially determining the toxic properties of a chemical (called hazard identification in the risk-assessment model), primary reliance is placed on toxicity studies conducted in animals.

As discussed more fully in other parts of the report (Chapters 3 and 7; Appendix B), a diverse battery of animal studies in rodent and nonrodent mammalian species, involving high doses, large numbers of animals, and often lifetime or even multigenerational exposure, can powerfully elucidate the potential toxicity of a chemical. Toxicologists can then apply standard methods to extrapolate the results seen at high doses in animals to the much lower doses ordinarily experienced by humans. In addition to their power to detect a chemical's toxic properties, animal studies play a central role in chemical risk assessment and safety evaluation as a matter of necessity. It would be economically prohibitive, but more importantly, ethically unacceptable to conduct in humans the kind of large-scale, long-term, highly invasive toxicity tests that would be required to determine the full range of a chemical's acute and chronic toxic properties, as can be done in animal studies. Human studies, therefore, can provide assessment of only some of the toxicities apparent in animals.

This does not mean, however, that data from human studies have no role in chemical risk assessment and safety evaluation. Much has been learned about the toxic properties of chemicals through the study of accidental human exposures to them, such as may occur in industrial accidents or unintentional environmental releases, and through epidemiological and occupational exposure studies that do not involve the intentional dosing of people but rather the examination of the effects of chemical exposures that people experience in their daily lives. Because they do not involve the intentional dosing of people, such studies do not raise the ethical concerns associated with a conscious decision to recruit individuals into research as a means of gaining knowledge. Established scientific principles and guidance provided by EPA and other agencies favor the use of data from such studies in risk assessment and safety evaluation when the data are available and scientifically relevant (EPA, 1989: NRC, 1993).[4]

[4]See the World Health Organization's Human Data Initiative at www.who.int/pcs/emerg_site/hdi/hdi_descr.html.

BOX 1.2
Four-Step Process for Human Health Risk Assessment

Following is the National Research Council's (1983) recommended four-step process for human health risk assessment, which it described as the use of the factual base to define the health effects of exposure of individuals or populations to hazardous materials and situations:

Step One: Hazard Identification

The first step in the risk-assessment process is to identify potential health effects that may occur from different types of pesticide exposure. EPA considers the full spectrum of a pesticide's potential health effects. Generally, for human health risk assessments, many toxicity studies are conducted on animals by pesticide companies in independent laboratories and evaluated for acceptability by EPA scientists. In addition, epidemiologic and in vitro data can be used. EPA evaluates pesticides for a wide range of adverse effects, from eye and skin irritation to cancer and birth defects in laboratory animals. EPA also may consult the public literature or other sources of supporting information on any aspect of the chemical.

Step Two: Dose-Response Assessment

The amount of a substance a person is exposed to is as important as how toxic the chemical might be. Dose-response assessment involves considering the dose levels at which adverse effects were or were not observed in test animals. In some cases, such as in the evaluation of potential carcinogens, dose-response information is used to estimate possible effects at doses below those at which effects were actually observed in animal studies.

There are circumstances, however, under which animal models may not be adequate to determine the potential toxicity of a chemical and epidemiological studies cannot be conducted in a way that is scientifically relevant to the exposure that is of regulatory concern to EPA. In some cases, intentional dosing of humans may be the only way to obtain the data needed to set regulatory standards or to protect public health. EPA itself conducts air chamber studies in which research participants, some with asthma or other conditions that make them vulnerable to air pollutants, are intentionally exposed to hazardous chemicals under controlled conditions designed to mimic or even exaggerate the "real world" circumstances in which the pollutants might be expected to cause symptoms. Data from such studies have played an important role in EPA's

Step Three: Exposure Assessment

People can be exposed to pesticides in three ways:

- Inhaling pesticides (inhalation exposure);
- Absorbing pesticides through the skin (dermal exposure); and
- Getting pesticides in their mouth or digestive tract (oral exposure).

The first task of an exposure assessment is to determine the concentration of the chemical to which humans are exposed, either through direct measurement or by estimate. Exposure assessment in an occupational setting consists of estimating long-term airborne exposures in the workplace. In the community, ambient concentrations of chemicals can be estimated from emission rates if transport and conversion processes are known. Assessments of exposure can be complicated by variations and personal habits among the population being studied and variable susceptibility.

Step Four: Risk Characterization

Risk characterization is the final step in assessing human health risks from pesticides. It is the process of combining the hazard, dose-response, and exposure assessments to describe the overall risk from a pesticide. It explains the assumptions used in assessing exposure as well as the uncertainties that are built into the dose-response assessment. The strength of the overall database is considered, and broad conclusions are made. EPA's role is to evaluate both toxicity and exposure and to determine the risk associated with use of the pesticide. Thus, the risk to human health from pesticide exposure depends on both the toxicity of the pesticide and the likelihood of people coming into contact with it.

ability to set standards to protect public health. However, such studies also raise important ethical and participant protection concerns.

In some cases, intentional human dosing studies can contribute to the process of extrapolating from animal results to estimate risks in humans, determine the mechanism by which a chemical affects human health, or determine the level in humans at which exposure to a chemical can be judged safe. These extrapolations ordinarily require that certain assumptions be made about the relationship between animal test results and what can be expected to occur in humans. As explained in Chapter 4, the replacement of these "default" assumptions with human data that more accurately reveal the likely human response can produce a more accurate risk assessment or safety evaluation.

The most germane example of replacing default assumptions with human data involves the use of human data in the safety evaluation of pesticide residues in food. In most cases, EPA determines safe levels of human exposure—levels at which there is "a reasonable certainty of no harm"—through a process that relies primarily on animal toxicity data and that involves the use of "uncertainty factors" to establish a safe level of exposure or Reference Dose (RfD) for the chemical. This process is described in detail in Chapter 7. It begins by identifying in the most sensitive animal model the highest dose at which no adverse effect from exposure to the chemical can be observed. This "no observed adverse effect level" (or NOAEL) is then divided by factors to account for the inherent uncertainty in extrapolating from animals to humans. Traditionally, EPA has applied 1 factor of 10 (the "interspecies" factor) to account for the possibility that a chemical is more toxic to humans than to the most sensitive animal species tested; and a second factor of 10 (the "intraspecies" factor) is applied to account for the possibility that humans vary widely in their response to the chemical, with some individuals being significantly more sensitive than others. Thus, under this approach to safety evaluation the animal NOAEL is divided by 100 to produce the RfD, or the dose that is judged safe for human consumption. If human data were available to demonstrate that humans were either substantially more or substantially less sensitive to the chemical than assumed by the 10-fold interspecies uncertainty factor, the factor could be adjusted upward or downward and thereby produce a scientifically more accurate RfD and safety evaluation. This possibility and the pesticide industry's response to it are what prompted this report.

EVENTS IN EPA'S PESTICIDE PROGRAM THAT PROMPTED THIS STUDY

In 1996, Congress passed the Food Quality Protection Act (FQPA), which substantially amended both the basic pesticide law, FIFRA, and the tolerance setting provisions of FFDCA. FQPA, which was the culmination of nearly two decades of debate over the efficiency and protectiveness of EPA's pesticide program, brought about important change. It mandated a single, health-based standard for all pesticides in all foods; provided special protections for infants and children; expedited approval of safer pesticides; created incentives for the development and maintenance of effective crop protection tools for American farmers; and required periodic reevaluation of pesticide registrations and tolerances to ensure that the scientific data supporting pesticide registrations will remain up-to-date in the future.

A significant change resulting from FQPA involved special protection for infants and children. The act mandated that EPA, in calculating safe levels of exposure for purposes of setting tolerances, apply an additional 10-fold safety factor for children, in addition to the interspecies and intraspecies factors ordinarily used, with the intention of taking "into account potential pre-and postnatal toxicity and completeness of the data with respect to exposure and toxicity" (FFDCA 408(b) (2) (C)).

This new requirement was based on a recommendation by the National Research Council (NRC) Committee on Pesticides in the Diet of Infants and Children, which, beginning in the late 1980s, worked for five years to examine regulatory and risk-assessment practices, assess dietary intake information, evaluate pesticide residue data, identify toxicological issues of greatest concern, and develop recommendations on policy changes and research needs. The committee concluded that the science on the susceptibility of children to pesticides was not very advanced and that children could be more or less sensitive than adults, depending on the chemical and the health endpoint of concern. One of the primary concerns of the committee, however, was that the developing organ systems in infants and children (e.g., nervous, endocrine, immune) might be particularly susceptible to some pesticides. In light of the lack of studies employing sensitive measures of such developmental effects, the NRC committee recommended that a third safety factor of up to 10 be applied (NRC, 1993). This and other committee recommendations were incorporated into the 1996 FQPA.

When this new FQPA safety factor is included with the interspecies and intraspecies uncertainty factors, the result is that the NOAEL typically is divided by 1,000 to yield a presumed safe level of exposure for purposes of setting tolerances. Any one of the 10-fold factors may be modified, however, on the basis of additional data demonstrating that a different safety factor is scientifically more valid—that is, more likely to produce an accurate expression of the safe dose. In the case of the FQPA 10-fold factor for children, information demonstrating that developing animals or children are not more sensitive to the pesticide being assessed than adults or that developmental toxicity is not the most sensitive endpoint could, in some cases, be used to support a safety factor of, for example, 3 or 1, instead of 10.

The interspecies uncertainty factor also can be modified for several reasons. If, for example, pharmacokinetic data are available to demonstrate that a substance's active metabolite is generated to a significantly different extent in laboratory animals than in humans, the standard 10-fold interspecies uncertainty factor could be replaced with a more specific (larger or smaller) interspecies factor. Similarly, evidence from pharmacodynamic studies showing that humans are more or less sensitive than animals to a relevant toxicity endpoint also could lead to the selection of a different interspecies uncertainty factor.[5] The interspecies uncertainty factor could potentially be set at 1 in cases where persuasive human data showed the sensitivity of humans and animals to the toxic endpoint of concern to be the same.

FQPA's requirement of an expedited reevaluation of older pesticides and the application of a third 10-fold safety factor apparently triggered concern on the part of some pesticide companies that certain commercially valuable pesticides would no longer meet the standards for food use. The evidence for this is that soon after enactment of FQPA, companies began submitting to EPA studies in humans that were intended to demonstrate that for certain chemicals the 10-fold interspecies uncertainty factor could be reduced or eliminated. As indicated by Table 1.1, of the 19 human studies submitted to EPA's pesticide program since 1991, 17 were submitted immediately following FQPA. If the studies and the reasoning behind them were accepted by EPA, they could have the effect of at least partially offsetting FQPA's new safety factor for children (by reducing the other safety factors) and increasing the likelihood that existing tolerances, and thus markets, for the pesticides would be maintained. The pesticides most commonly studied in these human experiments were cholinesterase-inhibiting organophosphates and carbamates, the two categories of pesticides that have been the subject of the most heated debate in the United States during their reevaluation.

Most of the human studies submitted since the enactment of FQPA are intended to establish a NOEL (no observed effect level) and a LOEL (lowest observed effect level). Such studies involve doses capable of eliciting a biological effect that is either potentially adverse in its own right or is considered a biomarker of exposure to a toxic agent, thus identifying the dose at which such effects can no longer be detected.

Some advocacy, scientific, medical, and environmental groups have objected to the industry's submission of the human studies, arguing that pesticide companies were subjecting research participants to potential risks in order to offset FQPA's tighter limits on pesticide exposure for children.[6] Some have expressed concern that the FQPA requirement for

---

[5]Some toxicologists use the term "pharmaco" when describing low-dose effects and "toxico" when describing high-dose effects or to differentiate between studies of drugs versus nontherapeutic chemicals. In this report, the committee chose to use the terms "pharmacokinetic" and "pharmacodynamic" to refer generally to kinetics and dynamics studies, rather than considering in each case whether an effect is, for example, toxic, or merely has an effect on a biomarker.

[6]Sass, J. National Resource Defense Council. 2003. Presentation at Public Forum: Providing Input to the Committee on the Use of Third Party Toxicity Research with Human Research Participants, January 8, 2003, National Academy of Sciences, Washington, D.C.

TABLE 1.1 Relevant Pesticide Studies Received by EPA Since April 1991

Year	Study
1992	A Safety and Tolerability Study of Aldicarb at Various Dose Levels in Healthy Male and Female Volunteers
1992	Amitraz: Report of a Double Blind Tolerance Study of Amitraz in Six Adult Healthy Volunteers
1997	Dichlorvos: A Single Blind, Placebo Controlled Randomized Study to Investigate the Effects of Multiple Oral Dosing on Erythrocyte Cholinesterase Inhibition in Healthy Male Volunteers
1997	Dichlorvos: A Study to Investigate the Effect of a Single Oral Dose on Erythrocyte Cholinesterase Inhibition in Healthy Male Volunteers
1997	Dichlorvos: A Study to Investigate Erythrocyte Cholinesterase Inhibition Following Oral Administration in Healthy Male Volunteers
1997	A Randomized Double Blind Ascending Dose Study to Determine the Safety and Tolerability of RH-7988 and to Establish a No Adverse Effect Level in Healthy Male Volunteers
1997	Safety and Tolerability Study of FCR 1272 [cyfluthrin]
1998	Amitraz: Human Volunteer Double-Blind Dermal Tolerance Study
1998	A Randomized Double Blind Ascending Single Oral Dose Study with Azinphos-methyl to Determine the No-Effect Level on Plasma and RBC Cholinesterase
1998	A Randomized Double Blind Ascending Oral Dose Study with Methomyl to Establish a No Adverse Effect Level
1998	ZA1296: Investigation of Systemic Exposure Following a Single Dermal Application of Spray Formulations to Healthy Male Volunteers
1998	Tolerance Study in Novartis Managers upon Repeated Oral Administration of Diazinon
1999	A Randomized Double Blind Ascending Oral Dose Study with Oxamyl
1999	A Rising Dose Toxicology Study to Determine the No-Observable-Effect Levels for Erythrocyte Acetylcholinesterase (AChE) Inhibition and Cholinergic Signs and Symptoms of Chlorpyrifos at Three Dose Levels
2000	A Rising Dose Toxicology Study to Determine the No-Observable-Effect Levels for Erythrocyte Acetylcholinesterase (AChE) Inhibition and Cholinergic Signs and Symptoms of Chlorpyrifos at Three Dose Levels
2000	A Randomized, Double-Blind, Ascending, Acute, Oral Dose Study of Diazinon to Determine the No Effect Level for Plasma and RBC Cholinesterase Activity in Normal, Healthy, Volunteers—Part A: Clinical Phase
2000	A Randomized, Double-Blind, Ascending, Acute, Oral Dose Study of Diazinon to Determine the No Effect Level for Plasma and RBC Cholinesterase Activity in Normal, Healthy, Volunteers—Part B: Analysis of DETP in Urine
2000	A Randomized, Double-Blind, Ascending, Acute, Oral Dose Study of Diazinon to Determine the No Effect Level for Plasma and RBC Cholinesterase Activity in Normal, Healthy, Volunteers—Part C: Analysis of Diazinon in Blood and G-27550 in Urine

an additional safety factor may have unintentionally created an incentive to test pesticides in humans (Gorovitz and Robertson, 2000). They also suggest that the pesticide manufacturers have an inherent conflict of interest because the research results would allow sustained or increased pesticide sales. This conflict, they argue, requires that there be a disinterested review of the validity of the data and the ethical acceptability of the research. Many opponents of this research argue that it is not acceptable to conduct this type of research under any circumstances.[7]

In 1998, the Environmental Working Group, a not-for-profit environmental research organization, published *The English Patients*, a report critical of EPA's practice of accepting data from third-party studies that intentionally expose people to pesticides for the purpose of determining safe or acceptable levels. The report recommended that EPA conduct a comprehensive review of past and current human experimentation; that it impose a moratorium on human experimentation for the purposes of pesticide registration until the review was completed; and that following the review, EPA adopt a policy to apply to studies conducted for the agency's regulatory programs such as the Federal Policy for the Protection of Human Subjects (the Common Rule), the ethical framework for human studies that many federal agencies, including EPA, apply to their own human research (see Chapter 2) (Environmental Working Group, 1998).

In response to growing public concern about these tests, industry representatives and some in the scientific community argued that human studies are necessary because they provide better data—i.e., animals are not always reasonable or accurate surrogates for humans—and that such studies are not very different from Phase 1 drug trials in which participants are exposed to potentially toxic drugs that offer them little if any prospect for benefit. Some asserted that human research in this area can advance the interests of public health within strict constraints and should not be abandoned, but rather refined and improved (McConnell, 2001).

In 1998, EPA announced that it was not relying on the submitted human pesticide studies to support decisions under FQPA. On July 27, 1998, the agency issued the following statement:

> EPA is deeply concerned that some pesticide manufacturers seem to be engaging in health-effects studies on human subjects as a way to avoid more protective results from animal tests under the new Food Quality

[7]Sharav, V. H. Alliance for Human Research Protection. 2003. Presentation at Public Forum: Providing Input to the Committee on the Use of Third Party Toxicity Research with Human Research Participants, January 8, 2003, National Academy of Sciences, Washington, D.C. Also available at www.ahrp.org/testimonypresentations/EPApesticide.html.

Protection Act. The government has in place very stringent standards that apply to federally funded research to ensure the protection of human subjects. EPA will be asking its independent Science Advisory Board to apply these same standards to pesticide data submitted to EPA by companies for review. No human test data has been used by EPA for any final decisions about acceptable levels of pesticide under the new food safety law. The protection of public health from adverse effects of pesticides can be achieved through reliance on animal testing and use of the highest ethical standards.[8]

PRIOR EPA ADVISORY PANEL REVIEW

EPA convened a Joint Subcommittee of its Science Advisory Board and the FIFRA Scientific Advisory Panel in 1998 to provide advice and comment to the agency on the scientific and ethical questions that had been raised about the use of data from intentional human dosing studies in making pesticide registration and tolerance decisions. The subcommittee was asked to address the value of such human studies and identify factors for consideration when (1) determining what constitutes an appropriate human study for use in environmental decision making; (2) making a judgment on what constitutes an ethically appropriate human study; and (3) determining if a study is appropriate (or inappropriate) for use. The agency also asked the subcommittee to discuss the risks and benefits of these studies for the research participants and for society, as well as the issues relevant to determining whether studies are in compliance with accepted ethical guidelines.

All but two of the subcommittee members could envision particular circumstances under which intentional dosing of research volunteers with small amounts of pesticides could be scientifically and ethically acceptable, subject to limitations described as ranging from "rigorous" to "severe." However, the majority also concluded that the information sought must not be available through other sources (e.g., animal studies and models or the study of incidental exposures) and that the information expected to be gained must promise reasonable health benefits to the exposed individuals or society at large.

A majority of the members of the subcommittee agreed to several basic findings and recommendations. These recommendations reemphasized the importance of protecting research participants and emphasized the need to establish a very high threshold of justification for studies that intentionally expose humans to toxic substances. Moreover, the recom-

---

[8]Available at www.epa.gov/scipoly/sap/1998/december/epastmt.htm.

mendations reflect the subcommittee agreement that justification of the use of humans in pesticide testing cannot be based on the ability to facilitate the interests of industry or of agriculture, but only to better safeguard the public health.

The EPA joint subcommittee expressed the need for scientifically rigorous protocols and stated that investigators should recognize that unintended exposures also provide valuable opportunities for research; thus, when possible, entities should take full advantage of the opportunity to gather information through careful incident follow-up rather than through intentional dosing studies. The subcommittee warned about the possible involvement in dosing studies of participants who are less than fully informed, the exposure of large numbers of participants to toxins, and the potential for skewed use of testing protocols in developing countries. It emphasized the need for EPA to adopt policies that reflect special concern for the interests of vulnerable populations, such as fetuses, children, adolescents, pregnant women, the elderly, and those with fragile health due to compromised respiratory function or other reasons. It recommended that in no case should developing humans (i.e., fetuses, infants, young children, or adolescents) be exposed to neurotoxic chemicals.

At a policy level, the EPA committee recommended that (1) EPA take whatever administrative action was necessary to extend the protections of the Common Rule (40 CFR Part 26) to all human research activities resulting in data submissions to the agency, including review by an Institutional Review Board (IRB) in compliance with the Common Rule; (2) the structure, function, and activities of EPA's IRBs as well as the external IRBs of entities submitting data should be under "active and aggressive scrutiny by EPA"; (3) EPA establish an internal ethics review organization for compliance oversight; and (4) data derived prior to the enactment of P.L. 92-516 (amendments to FIFRA) need not be rejected, even if the research was conducted unethically.

In a minority report, several committee members argued that the majority report underplayed risks to humans from intentional experimental dosing and that human studies as currently designed fail to provide information about safe levels of intake of pesticides by humans, especially children. They further argued that the majority recommendations would lead to more intentional dosing studies and eventually higher levels of pesticide exposures in the U.S. population.

Following the submission of these recommendations, EPA concluded that scientific and ethical questions remained and that the issues raised could be just as relevant to many of the agency's other programs, citing EPA's past reliance on data from intentional human dosing studies in decision making regarding particulate and ozone air pollution.

On December 14, 2001, EPA suspended the use of data from chemical safety studies in humans pending the completion of a report by the National Academy of Sciences (NAS) on the scientific validity and ethical acceptability of human studies of pesticides and other substances conducted by clinical laboratories under contract to private companies.[9] This report is the result of that decision. Before describing the charge EPA gave to this NAS committee—the Committee on the Use of Third Party Toxicity Research with Human Research Participants—it is important to clarify EPA's current policy concerning the ethical oversight of human studies.

EPA'S POLICY REGARDING ETHICAL OVERSIGHT OF HUMAN STUDIES

In administering its chemical regulatory statutes, EPA conducts and sponsors a wide variety of research studies involving humans, including observational studies of everyday common exposures, epidemiological studies, and deliberate dosing studies. EPA is a signatory to Subpart A of the Federal Policy for the Protection of Human Subjects (the Common Rule), the requirements of which are described in further detail in Chapter 2 and summarized in Box 1.3. The Common Rule is codified in the regulations of the Department of Health and Human Services (DHHS) at 45 CFR 46 and adopted for EPA purposes in 40 CFR Part 26.[10] Research conducted or sponsored by EPA must be in compliance with the Common Rule.

---

[9]Following the announcement, CropLife America, AMVAC Chemical Corporation, and Aventis CropScience USA LP petitioned the U.S. Court of Appeals for the District of Columbia stating that the moratorium constitutes an unlawful de facto regulation. Moreover, the petitioners claimed that the moratorium contravenes the clear requirement of the FFDCA that EPA consider all relevant reliable data in making pesticide decisions (21 U.S.C. §346a(b)(2)(D)), and the provision of FIFRA recognizing that human clinical studies are not invalid when people "freely volunteer to participate in the test," and when they "are fully informed of the nature and purposes of the test," and of any reasonably foreseeable health consequences (7 U.S.C. §136j(a)(2)(P)). Oral arguments were heard March 17, 2003, and on June 3, 2003, the District of Columbia Circuit Court invalidated EPA's directive suspending reliance on third-party human studies on the grounds that EPA had failed to follow correct procedures in suspending use of such studies. The court did not render any substantive judgment about the EPA action.

[10]Each signatory to the Common Rule promulgated the same set of regulations within its statutory authority. The original regulations, as codified by the Department of Health and Human Services, are found at 45 CFR 46. EPA promulgated the regulations as 40 CFR 26. This report will refer to the regulations as codified by EPA in 40 CFR 26, which includes only Subpart A. EPA has not yet signed on to Subparts B through D, which focus on protections for vulnerable subjects, such as children, pregnant women, and prisoners.

BOX 1.3
Common Rule Requirements for IRB Review

In essence, the Common Rule has two major substantive requirements—that human research receives review by an independent body for its ethical acceptability and that the IRB determine the requirements for informed consent by potential research participants.

Current regulations instruct IRBs in approving research studies as follows:

In order to approve research...the IRB shall determine that all of the following requirements are satisfied:

1. Risks to subjects are minimized: (i) by using procedures which are consistent with sound research design and which do not unnecessarily expose subjects to risk, and (ii) whenever appropriate, by using procedures already being performed on the subjects for diagnostic or treatment purposes.

2. Risks to subjects are reasonable in relation to anticipated benefits, if any, to subjects, and the importance of the knowledge that may reasonably be expected to result. In evaluating risks and benefits, the IRB should consider only those risks and benefits that may result from the research (as distinguished from risks and benefits of therapies subjects would receive

In July 1999, EPA issued a directive clarifying and extending its policies for application of the Common Rule, making it applicable to a broader range of human studies, including research that (1) involves the gathering of physiological measurements (e.g., monitoring a subject's cardio-respiratory performance) or the collection of body fluids, tissue, or expired air from subjects; (2) requires subjects to perform specific tasks other than their normal activities or to manipulate their environment (i.e., to modify their exposure); or (3) gathers or records private information (as defined in 40 CFR 26.102 (f)(2)) in a manner that associates such information with an identifiable subject.

Before EPA initiates research involving humans in one of its own laboratories or supports such research, such as through a contract, grant, cooperative agreement or interagency agreement, the study must be approved by the EPA Human Subjects Research Review Official (the Review Official) or be determined by the Review Official to be "exempt research," according to the exemptions provided in the regulations at 40 CFR 26.101(b). To obtain approval by the Review Official, the agency official even if not participating in the research). The IRB should not consider possible long-range effects of applying knowledge gained in the research (for example, the possible effects of the research on public policy) as among those research risks that fall within the purview of its responsibility.

3. Selection of subjects is equitable. In making this assessment the IRB should take into account the purposes of the research and the setting in which the research will be conducted and should be particularly cognizant of the special problems of research involving vulnerable populations, such as children, prisoners, pregnant women, mentally disabled persons, or economically or educationally disadvantaged persons.

4. Informed consent will be sought from each prospective subject or the subject's legally authorized representative, in accordance with, and to the extent required by 40 CFR 26.116.

5. Informed consent will be appropriately documented, in accordance with, and to the extent required by 40 CFR 26.117.

6. When appropriate, the research plan makes adequate provision for monitoring the data collected to ensure the safety of subjects.

7. When appropriate, there are adequate provisions to protect the privacy of subjects and to maintain the confidentiality of data.

8. When some or all of the subjects are likely to be vulnerable to coercion or undue influence, such as children, prisoners, pregnant women, mentally disabled persons, or economically or educationally disadvantaged persons, additional safeguards have been included in the study to protect the rights and welfare of these subjects (40 CFR 26.111).

responsible for the research must submit to the Review Official documentation showing to the satisfaction of the Review Official that the research will be conducted in accordance with the Common Rule. The Review Official may withhold approval of any proposal that does not adequately protect the rights and welfare of the participants.

If EPA-funded studies are conducted at a non-EPA site, the study must be reviewed by the research institution's IRB, and the research site should have an assurance of compliance on file with either EPA or the Office for Human Research Protections within DHHS. If the study is conducted at an EPA facility, EPA requires that the research be reviewed by an appropriate IRB, and a Review Official, who is a member of EPA's Office of Research and Development, ensures that IRB review has occurred.

With respect to studies conducted for EPA regulatory purposes by third parties, such as the pesticide studies discussed earlier, the Common Rule does not necessarily apply, and there is no established system within

EPA for reviewing the conduct of such studies from an ethical and participant protection perspective.

CHARGE TO THE COMMITTEE

EPA asked NAS to conduct a review of the complex ethical and scientific issues posed by EPA's possible use of third-party studies that intentionally dose humans with toxicants to identify or quantify their effects, including specifically studies to ascertain a NOAEL. The specific tasks assigned to the committee by EPA are provided in Box 1.4. Although the recent controversies about EPA's use of human research studies have primarily involved the pesticide program, the agency's request to NAS was not limited to such studies, but rather asked for advice about third-party studies generally. Accordingly, the committee reviewed the question of third-party studies generally, always bearing in mind that the pesticide studies have been a particular object of concern both inside and outside the agency.

The tasks EPA assigned to this committee require consideration of some difficult issues, but they do not require the committee to invent the basic standards that govern human research in the United States. These standards are already embodied in the Common Rule and in other authoritative statements of principle on the ethical conduct of human research (as discussed in Chapter 2). Rather, the committee's assigned task was to consider how those standards should be applied in the particular case of intentional human dosing studies conducted by third parties for EPA regulatory purposes. The existing ethical standards are sufficiently general, however, and the studies in question are sufficiently different from most human research conducted in the United States that applying the standards to the cases at hand requires considerable analysis. The committee carried that analysis far enough to provide guidance to EPA on how, as a general matter, the standards should be applied. This analysis includes careful review by the committee and committee staff of some of the specific human studies that have been submitted to EPA. It was not the committee's charge, however, to make decisions or recommendations on the ultimate acceptability of any specific study for regulatory purposes. This requires a depth of analysis, both ethical and scientific, that is beyond the scope of the committee's charge.

As noted earlier, the scientific validity of a human study for a particular regulatory purpose is intertwined with the study's ethical acceptability. The committee explored in great depth principles of both ethical and scientific validity in order to make recommendations about how accepted principles should be applied here. The committee also accepted the charge of considering the scientific basis that an otherwise ethically sound human study could rely on to address one particular regulatory issue: the possible alteration of the interspecies uncertainty factor (discussed in Chapter 7). This is an important and sensitive science policy issue that lies behind much of the public interest and controversy surrounding the issues considered by this committee.

Finally, although the committee's charge was directed to third-party human studies, the committee noted that the ethical and scientific issues are fundamentally the same regardless of whether a human study is conducted by a third party or by EPA and that the same basic ethical framework should apply to both third-party and non-third-party studies. The ethical issues of concern about third-party studies arise because they potentially impose health risks on human beings, and with regard to this characteristic, third-party studies and agency-sponsored studies are indistinguishable. If third-party and agency-sponsored studies should be treated differently within an ethical framework, it would have to be because they differed systematically with regard to some other characteristic, such as the benefits to be derived from the studies or the ability of subjects to provide informed consent. Such a conclusion, however, would have to emerge from the application of the basic framework to specific experiments, rather than a priori, as an operating assumption.[11] For this reason, the committee's recommendations apply to both third-party and EPA-supported studies—that is, to any research sponsor submitting human data to EPA for regulatory purposes.

NATIONAL ACADEMY OF SCIENCES COMMITTEE PROCESS

To conduct this study, a committee composed of members with expertise in ethics, law, pharmacology, toxicology, genetics, pediatrics, statistics/biostatistics, economics, epidemiology, risk assessment, and clinical trials was established under the auspices of the National Academies' Science, Technology, and Law Program and in accordance with NAS procedures and policies regarding the nomination and appointment of study committees. The names and biographies of the nominated individuals were posted to the NAS website for public review and comment. Comments were considered during the committee bias and composition discussion at the first meeting on December 16, 2002.

[11]The committee did find some important distinctions to be drawn between agency-sponsored and third-party studies, primarily related to the processes for ensuring compliance with ethical standards.

BOX 1.4
Specific Statement of Task

According to EPA's charge to NAS, the scope of information gathered by the committee and the topics on which public input shall be solicited include, but are not necessarily limited to:

1) Whether and if so to what extent EPA's decision to accept, consider, or rely on a third-party, human toxicity study should depend on:

a) whether the study was conducted in substantial compliance with the provisions of the Common Rule or another standard for the protection of human subjects;

b) the type of substance tested (e.g., pharmaceutical, pesticide, environmental contaminant);

c) whether the results of the study tend to indicate that the substance tested is more risky or less risky than is indicated by other available data;

d) the statistical power of the study, or the ability or inability to measure the same endpoints in humans that have been observed in animal testing of the same substance or other specific characteristics of the study design;

e) when the study was conducted in relation to the date of any statement of policy by EPA regarding the ethical conduct of such studies;

f) whether there are alternative methods of obtaining data of comparable scientific merit that would not involve deliberate dosing of human subjects;

g) the nature of the test sponsor's interest in a regulatory matter that could be affected by consideration of the data;

h) how EPA intends to use the results in its regulatory decision making (e.g., to reduce or remove the traditional 10-fold interspecies uncertainty factor, or to provide an endpoint for use in calculating a reference dose for the test substance, or for some other purpose);

i) whether the study has been submitted in response to a regulatory requirement of EPA, or whether it was conducted in conformity with an EPA Guideline;

j) EPA's assessment of the actual or potential benefits, if any, to the individual human subjects of the research, or to society;

2) Under what circumstance(s), if any, the availability of human data should lead EPA to consider reducing or removing the customary 10-fold interspecies uncertainty factor;

3) What existing standards (e.g., the Common Rule, the *Declaration of Helsinki*) are available for evaluating the design and the conduct of research with human subjects, and which of these standards would be most appropriate in judging whether human toxicity studies submitted to EPA in

support of a regulatory decision were conducted ethically and in a way fully protective of the interests at safety of the human subjects;

4) Whether and if so how the requirements of the Common Rule should be extended to the conduct of third-party research with human subjects intended for submission to EPA in support of a regulatory decision; and

5) To what extent and how the submitter of research with human subjects to EPA should be required to document or otherwise demonstrate compliance with appropriate standards for the protection of human research subjects—e.g., fully informed and fully voluntary participation and independent oversight of research design and conduct by an Institutional Review Board.

The committee report shall address the full range of relevant issues (including those listed above) and provide advice—including suggesting appropriate criteria and factors, and reasons for all recommendations—for EPA to consider in establishing agency policy. The committee report shall include, but is not limited to:

1) Identification of the committee members and a description of the process by which the committee conducted its business;

2) A description of the opportunities for the public to be informed of and to participate in the committee's process;

3) A description of the current legal (statutory or regulatory) requirements governing the conduct of third-party research with human subjects and consideration by a federal agency of the result of such research;

4) A description of the current policies and practices of other federal agencies regarding acceptance, consideration, and reliance on third-party research with human subjects and a description of any requirements and enforcement practices of such agencies relating to the ethical conduct of such research;

5) The views of the committee regarding the types of third-party human research, if any, that EPA should always refuse to accept, consider, or rely on;

6) The views of the committee regarding minimum standards relating to the protection of human subjects which should be met in the design and conduct of a study with human subjects, in order for EPA to accept, consider, and rely on the results of the study in regulatory decision making;

7) The views of the committee regarding the minimum scientific standards relating to the reliability and relevance of the results that should be met for a human study in order for EPA to accept, consider, and use the results of the study in regulatory decision making; and

8) The views of the committee regarding the best way(s) in which any minimum standard for the conduct of third-party research with human subjects should be imposed, implemented, and enforced.

The committee met 6 times over 12 months in open and closed meetings, convened numerous conference calls, and invited testimony from a number of individuals. In addition, it convened one public forum on January 8, 2003, to receive public input on the topics under consideration. The committee received and reviewed studies voluntarily submitted by a number of companies that had previously conducted intentional human dosing studies and that had submitted their results to EPA for consideration. Some of these submissions included complete files on a particular chemical, while others were partial files. All of these materials were placed in the National Academies' public access file for this project. In addition, committee staff filed a freedom of information request with EPA for all information relevant to the intentional human dosing studies that had been submitted to the EPA Office of Pesticide Programs. Committee staff reviewed these studies and briefed the committee on their findings. In addition, EPA provided the committee with a copy of all of the public comments submitted to the agency in response to an Advance Notice of Proposed Rulemaking concerning intentional dosing studies that it published on May 7, 2003 (EPA, 2003).

ORGANIZATION OF THE REPORT

Determining the organization of this report has been a challenge, because the issues and analysis involved are so intertwined. Although an effort has been made to provide a coherent narrative, it has been necessary to make numerous cross-references among chapters.

Chapter 2 expands upon this chapter's discussion of the Common Rule and the general ethical and regulatory framework for the oversight of research involving humans. Chapter 3 describes the relevant types of intentional human dosing studies and recommends criteria for assessing the scientific validity of such studies for ethical purposes and for EPA regulatory decision-making purposes. This includes consideration of the scientific justification for conducting a study and issues of study design and reporting.

Chapter 4 provides a framework for assessing and balancing the risks and potential benefits of intentional dosing studies in humans and provides examples of how this framework might apply to human studies of the kind that have been submitted to EPA. Chapter 5 focuses on additional ethical considerations in the conduct of human studies, such as the selection of research participants, payment for participation in research, informed consent, compensation for research-related injuries, and the use of results from ethically problematic studies. It also outlines the committee's conclusions and recommendations to EPA regarding procedures for the ethical review of toxicant studies, before and after they are conducted.

Chapter 6 recommends a procedural framework for EPA's implementation of the scientific and ethical principles described in earlier chapters, including the formation of a new review board within EPA. Chapter 7 provides the committee's discussion of and recommendations for EPA's use of human study results in risk assessment and in considering possible adjustments in the interspecies uncertainty factor.

REFERENCES

Environmental Protection Agency (EPA). 1989. *Risk Assessment Guidance for Superfund, Vol. 1: Human Health Evaluation Manual,* Interim Final. EPA/540-1-89/002, December. Available at www.epa.gov/cgi-bin/claritgw?op-display&document=clserv:OSWER:1175; &rank=4&template=epa).

EPA. 2003. Human Testing; Advanced Notice of Proposed Rulemaking. *Federal Register* 68:24410-24416.

Environmental Working Group. 1998. *The English Patients: Human Experiments and Pesticide Policy.* Washington, D.C.: Environmental Working Group. Available at www.ewg.org/reports/english/English.pdf.

Gorovitz, S., and H. Robertson. 2000. Pesticide toxicity, human subjects, and the Environmental Protection Agency's dilemma. *Journal of Contemporary Health Law Policy* 16(2):427-458.

McConnell, E. 2001. The value of human testing of pesticides. *Human and Ecological Risk Assessment* 7(6):1575-1581.

National Bioethics Advisory Commission (NBAC). 2001. *Ethical and Policy Issues in Research Involving Human Participants: Vol. 1.* Bethesda, MD: U.S. Government Printing Office.

National Research Council (NRC). 1983. *Risk Assessment in the Federal Government: Managing the Process.* Washington, D.C.: National Academy Press.

NRC. 1993. *Pesticides in the Diets of Infants and Children.* Washington, D.C.: National Academy Press.

2

The Regulatory Framework for Protecting Humans in Research

INTRODUCTION

Through federal regulations, the U.S. government has established a system of protections for research participants. Eighteen federal agencies and departments adhere to the Federal Policy for the Protection of Human Subjects, or the Common Rule (45 CFR 46),[1] which is a set of identical regulations codified by each agency. This system of protections, however, applies only to research that is conducted or funded by an agency that is subject to the Common Rule or that is subject to Food and Drug Administration (FDA) review and approval. Many institutions hold assurances of compliance to the Common Rule, which are negotiated with the federal government. Such assurances cover all of the institution's research involving humans that is conducted or supported by one of the federal departments or agencies that have adopted the Federal Policy.

In considering the appropriate oversight of third-party human research conducted for Environmental Protection Agency (EPA) regulatory purposes, it is useful to understand the development of the system of protections to which EPA must adhere under the Common Rule, as well as the practices of other federal agencies in this regard, as lessons learned from the past and in other research contexts can inform the development

[1]Each signatory to the Common Rule promulgated the same set of regulations within its statutory authority. EPA promulgated the regulations as 40 CFR 26. This report will refer to the regulations as codified by EPA in 40 CFR 26, which includes only Subpart A. EPA has not yet signed on to Subparts B through D, which focus on protections for vulnerable subjects, such as children, pregnant women, and prisoners.

and improvement of EPA regulatory policy for third-party studies. Of note, EPA previously has not applied the Common Rule protections to privately sponsored (third-party) studies of regulated substances. Were EPA to include such studies in its oversight system, it would be useful to consider how those regulations might apply.

HISTORY OF THE DEVELOPMENT OF FEDERAL REGULATIONS

Public policies regarding the ethical treatment of humans in research began forming in the late 1940s, largely in response to atrocities committed by Nazi investigators who were tried before the Nuremberg Military Tribunal (*United States* v. *Karl Brandt et al.*).[2] In 1946, the American Medical Association adopted its first code of research ethics (AMA, 1946), which ultimately influenced the Nuremberg Tribunal's standards for ethical research (Moreno, 1999), embodied in the ten "basic principles" for human research, now known as the Nuremberg Code.

The first principle of the Nuremberg Code states that, "the voluntary consent of the human subject is absolutely essential." This absolute requirement reflects the code's origins in discussions about research with healthy individuals, particularly those who had no opportunity to refuse. According to the code, investigators alone are responsible for obtaining informed consent and deciding whether their research is in accord with the ethical principles.

Following the issuance of the Nuremberg Code, several federal agencies began establishing policies for human research. In 1953, Department of Defense Secretary Charles Wilson issued a directive outlining a policy for human research related to atomic, biological, and chemical warfare (Wilson, 1953). Wilson's policy included a prohibition on research involving prisoners of war and a requirement that the secretary of the appropriate military service approve human research studies. Also in 1953, the National Institutes of Health (NIH) Clinical Center established a policy requiring independent review of research and participants' written consent, at least for research involving patient volunteers and/or "unusual hazard" (NIH, 1953). In 1954, these dual protections of independent review and written informed consent were extended to all NIH intramural research involving "normal volunteers."

However, widespread adoption of ethical principles in the conduct of human studies was slow to develop. Some believed that the Nuremberg

[2]The "Medical Case," *United States* v. *Karl Brandt et al.*

Code was meant to apply only to research with healthy individuals and not to research with patients as participants. Moreover, U.S. policy makers were concerned about intruding into the doctor-patient relationship, and until national attention focused on some research scandals in the 1960s, specific human protections in that context seemed unnecessary.

In 1962, Congress passed the Kefauver-Harris amendments to the Federal Food, Drug, and Cosmetic Act. The amendments are best known for requiring FDA to evaluate new drugs for efficacy in addition to safety (P.L. 87-781). The amendments also required the informed consent of participants in the testing of investigational drugs, although permissible exemptions applied, and they emphasized the need for investigators to control the drug supply.

Then, a series of events began to focus attention on the need for closer regulation of human studies. In early 1964, newspapers began to report on an NIH-funded study at the Brooklyn Jewish Chronic Disease Hospital in which investigators had injected cancerous cells into elderly patients. The investigators claimed to have obtained informed consent from the study participants, but many were incapacitated or did not speak English, and those able to give consent were not told that the cells to be injected were cancerous (Faden and Beauchamp, 1986; Jonsen, 1998).

In 1966, Henry Beecher published a startling indictment of research practices in the United States, presenting 22 examples of "unethical or questionably ethical studies" published in major medical journals (Beecher, 1966). One of the studies described by Beecher was an investigation of hepatitis involving the injection of a mild strain of the virus into children at the time of their admission to the Willowbrook State School for the Retarded in New York. Parental consent had been obtained, but the consent form might have been misleading, and parents may have been unduly influenced by the fact that research participants were put at the top of a long waiting list for admission (Faden and Beauchamp, 1986).

In response to growing concerns about documented and alleged research abuses, NIH developed policies to force NIH units to take more responsibility for research ethics (Faden and Beauchamp, 1986). In 1966, the Public Health Service (PHS) issued a new policy for studies sponsored but not conducted by the agency, requiring independent review of research by a committee of the investigator's "institutional associates" (PHS, 1966). A memorandum accompanying the policy stated that a group of people from different disciplines, familiar with the investigator but "free to assess his judgment without placing in jeopardy their own goals," would be required for the review (Stewart, 1966). NIH initiated a system in which it negotiated assurances of compliance with the PHS policy from each institution receiving funding. As an enforcement measure, NIH could withhold funds.

NIH would later formally establish the Office for Protection from Research Risks (OPRR) in 1972 to implement and enforce these policies, and eventually this office—renamed the Office for Human Research Protections (OHRP) in 2000—assumed a lead role in the protection of research participants within the entire Department of Health and Human Services (DHHS).

Until 1966, the PHS Policy for Clinical Investigations with Human Subjects applied only to extramural research, and only to NIH grantees. In 1971, 5 years after the PHS policy was established, what was then the Department of Health, Education and Welfare (DHEW) developed more detailed guidance and justification for review committees in the form of the "Yellow Book" (DHEW, 1971).

Perhaps the most significant event to force the development and use of a more uniform and systematic approach to protecting research participants came in the aftermath of a 1972 *New York Times* article that reported the details of the Tuskegee Syphilis Study, sponsored by PHS since the early 1930s (Heller, 1972). Although a formal protocol never existed, the study aimed to trace the natural history of syphilis in poor African American males living in Macon County, Alabama. Participants were not told of the purpose of the study and were actually misled into believing that they were being treated for syphilis. Investigators continued the study even after penicillin became widely available and prescribed for the treatment of syphilis. In exchange for participation, the men received some unrelated health care services, free meals, and transportation, and later in the study a $50 burial stipend (Jones, 1981). A PHS investigation in 1973 found the study to be ethically unjustified, and it was halted. The surviving participants were offered treatment. In addition, a PHS advisory panel determined that existing procedures for protecting research participants were not adequate. The panel recommended that "Congress should establish a permanent body with the authority to regulate *at least* all Federally-supported research involving human subjects" (Tuskegee Syphilis Study Ad Hoc Advisory Panel, 1973).

In 1973, the Senate Labor and Public Welfare Committee began a series of hearings on human experimentation, which led to an agreement that DHEW would issue regulations governing research with humans (ACHRE, 1995). The resulting regulations were promulgated in May 1974 (DHEW, 1974) (21 CFR Part 50), and the National Research Act was signed in July of that year (P.L. 93-348). The National Research Act also established the National Commission for the Protection of Human Subjects of Biomedical and Behavioral Research (National Commission) to provide ethical and policy analysis related to human research. The National Commission is perhaps best known for its *Belmont Report: Ethical Principles and Guidelines for the Protection of Human Subjects of Research* (National Commission, 1979). This report identified three fundamental ethical principles applicable to research with humans—respect for persons, beneficence, and justice—which translated respectively into provisions for informed consent, assessment of risk and potential benefits, and selection of participants. For example, the application of the ethical principle of respect for persons gives rise to the concern that consent be properly obtained from fully informed participants and that special consideration be given to vulnerable persons who may lack the capacity to consent. The application of the principle of beneficence leads to the necessity of assessing and balancing risks and potential benefits. The principle of justice requires investigators to attend to the process of recruiting research participants, with particular attention to vulnerable populations. The National Commission also recommended that special regulations be adopted to protect children in research, which formed the basis of Subpart D of the Common Rule.

DHEW regulations already contained specific provisions for obtaining and documenting informed consent and guidance on assessing risk and benefit. The *Belmont Report* recommended that additional attention be given to the equitable selection of participants. In response to the *Belmont Report*, DHHS and FDA revised their regulations (45 CFR 46; 21 CFR 50, 56). The revised regulations placed primary emphasis on obtaining and documenting voluntary informed consent, but provided little guidance on assessment of risk and potential benefit or the selection of research participants.

In 1981, the President's Commission for the Study of Ethical Problems in Medicine and Biomedical and Behavioral Research (President's Commission) was established. In several reports the President's Commission examined the general structure and implementation of existing research protections (President's Commission, 1981; President's Commission, 1983). Its notable recommendations from its 1981 and 1983 reports include the following:

- All federal agencies should adopt the regulations of DHHS (45 CFR 46).
- Each federal agency should apply one set of rules consistently to all of its subunits and funding mechanisms.
- Principal Investigators should be required to submit annual data on the number of subjects in their research and the number and nature of adverse events.
- Federal agencies should clarify the meaning of certain procedural requirements of existing regulations, particularly what is meant by "Institutional Review Board (IRB) review."
- Federal agencies that do not already do so should, as soon as prac-

ticable, identify the IRBs responsible for the initial and continuing review of research for which they have regulatory authority.

• The prospective review of institutional assurances of compliance with applicable regulations should consider the amount and types of research that each IRB anticipates reviewing and should determine that requirements regarding IRB composition are met, that sound procedures have been established for the IRB's review of the research, and that the institution understands its responsibilities for protecting participants.

• A broad educational and monitoring program covering the protection of research participants and designed to reach investigators, IRB members, and research administrators should be conducted. Among the various activities included in the program should be site visits of research institutions using experienced IRB members and staff as site visitors.

The President's Commission also recommended, as did the National Commission, that special protections be codified for children. In response, DHHS promulgated regulations in 1983 governing research with children (Subpart D).

In response to the President's Commission's concern about the lack of standardization of regulations across federal agencies and departments, the White House convened an interagency ad hoc committee to develop what would become the Common Rule (the Federal Policy for the Protection of Human Subjects), a set of identical regulations codified by various agencies. The standardization process was slow, taking nearly 10 years to occur. In 1991, the regulations known as the Common Rule were simultaneously published in the *Federal Register* by 15 departments and agencies. The Office of Science and Technology Policy in the Executive Office of the President did not codify the Common Rule, even though it signed the Federal Policy, because it did not conduct or sponsor research (NBAC, 2001). The Common Rule also regulates research conducted or sponsored by two other federal agencies that are not signatories to the Common Rule but that are bound nonetheless through public law (the Social Security Administration [P.L. 103-296]) or by Executive Order (the Central Intelligence Agency [E.O. 12333]). Thus, the Common Rule has 15 codifications and 16 signatories, and it covers 18 federal agencies (see Table 2.1). The rule expanded the scope of regulated research and provided some standardization across departments, with DHHS, primarily through OPRR, playing a key role in its development.

THE COMMON RULE

The Common Rule applies to all research involving humans "conducted, supported or otherwise subject to regulation by any federal de-

TABLE 2.1 Federal Agencies Subject to the Common Rule

- Agency for International Development
- Central Intelligence Agency
- Consumer Product Safety Commission
- Department of Agriculture
- Department of Commerce
- Department of Defense
- Department of Education
- Department of Energy
- Department of Health and Human Services
- Department of Housing and Urban Development
- Department of Justice
- Department of Transportation
- Department of Veterans Affairs
- Environmental Protection Agency
- National Aeronautics and Space Administration
- National Science Foundation
- Office of Science and Technology Policy
- Social Security Administration

partment or agency which takes appropriate administrative action to make this policy applicable to such research." Thus, it specifically allows agencies with regulatory authority to apply the Common Rule to regulated research (40 CFR 26.101(a)).[3]

Even though the federal regulations cover a large portion of human research conducted domestically, and in some cases overseas, they are limited in their reach. In fact, if federal funds are not involved or if regulatory approval is not required, research activities involving humans might not be subject to any form of oversight. The regulations also do not apply to many areas of research funded and conducted by businesses, private nonprofit organizations, and state or local agencies, although such research is subject to federal regulation if it involves the development of medical devices or drugs requiring approval by the FDA or if it is con-

[3]DHHS does not require FDA to apply the Common Rule to the research FDA regulates. FDA also has its own regulatory authority over research involving food and color additives, investigational drugs for human use, medical devices for human use, biological products for human use being developed for marketing, and electronic products that emit radiation. FDA also regulates research intended to support a change in the labeling of marketed products. To this regulated research, FDA applies its own set of regulations (21 CFR 50, 56) that are generally, but not entirely, similar to the Common Rule. FDA is bound to DHHS regulations when it conducts its own research.

ducted at an institution that has voluntarily agreed to apply Common Rule requirements to all research it conducts (see the discussion of assurances below).

Moreover, the Common Rule did not create a shared mechanism for interpreting and implementing the regulations at the federal level. Some departments have not established offices for interpreting and implementing the regulations; in some cases, a single individual is responsible for oversight activities (NBAC, 2001). In 2001, the National Bioethics Advisory Commission (NBAC) found that departments and agencies bound to the Common Rule sometimes interpret the regulatory requirements differently.

Finally, the Common Rule has four subparts. Subpart A is the only part signed on to by all participating agencies. Subparts B through D address specific additional protections and considerations for research involving fetuses, pregnant women, and human in vitro fertilization (Subpart B), prisoners (Subpart C), and children (Subpart D). Only DHHS and the Department of Education are signatories to Subpart D, and only DHHS adheres to Subparts B and C. EPA has signed on to Subpart A only.

Nonetheless, there are basic concepts contained in the regulations that provide a framework and guidance for federal oversight, even though the specific policies and procedures adopted by a department or agency for implementation might differ.

Minimal Risk

Determining whether a study poses more than minimal risk is a central ethical and procedural function of the IRB as outlined in the federal regulations (40 CFR 26.102(i)). The regulations call for the classification of research as involving either minimal risk or greater than minimal risk. When used as a sorting mechanism, this classification determines the level of review required of an IRB. For example, under the current regulations, if a research study is determined to pose only minimal risk and involves a procedure contained on an expedited review list, it may be evaluated using the expedited review process in which the IRB chair or a designee may review the research study in accordance with all the required regulations (40 CFR 26.110(b)).

Research involving more than minimal risk requires full IRB review. As the risk of research increases above the minimal risk threshold, protections for participants become more stringent. For example, with greater than minimal risk research, the process of informed consent cannot be waived or altered (40 CFR 26.116(d)).

The language of the regulations, however, provides an ambiguous standard for minimal risk, under which risks involved in a research study are compared to those encountered in daily life. As defined in the federal regulations:

> Minimal risk means that the probability and magnitude of harm or discomfort anticipated in the research are not greater in and of themselves than those ordinarily encountered in daily life or during the performance of routine physical or psychological examinations or tests (40 CFR 26.102(i)).

It is unclear whether this applies to those risks found in the daily lives of healthy individuals or those of individuals who belong to the group targeted by the research. In 2001, NBAC recommended that IRBs use a standard related to the risks of daily life that are familiar to the general population for determining whether the level of risk is minimal or more than minimal, rather than using a standard that refers to the risks encountered by particular persons or groups. At present, minimal risk is most commonly applied to studies in which there is no pharmacologic intervention (e.g., epidemiological studies or studies in which drug blood levels are measured in people already receiving the drug for a therapeutic purpose). Venipuncture is generally considered a minimal risk. There are, however, many kinds of studies that would seem to involve a very small movement above minimal risk, such as most bioavailability studies of marketed drugs or very short studies of the effects of a usual dose of a drug on a biomarker (blood pressure, blood sugar). These sorts of risks are not extensively discussed, although the concept of "a minor increase over minimal risk" appears in the Subpart D of the Common Rule related to children.[4]

Institutional Review Board Approval of Research

The current regulations at 40 CFR 26.111 provide IRBs with the following instructions:

> In order to approve research . . . the IRB shall determine that all of the following requirements are satisfied:
>
> 1. Risks to subjects are minimized: (i) by using procedures which are consistent with sound research design and which do not unnecessarily expose subjects to risk, and (ii) whenever appropriate, by using procedures already being performed on the subjects for diagnostic or treatment purposes.

[4]See also the discussion in *Clarifying Specific Portion of 45 CFR 46 Subpart D That Governs Children's Research, Report from the National Human Research Protections Advisory Committee.* Available at ohrp.osophs.dhhs.gov/nhrpac/documents/nhrpac16.pdf.

2. Risks to subjects are reasonable in relation to anticipated benefits, if any, to subjects, and the importance of the knowledge that may reasonably be expected to result. In evaluating risks and benefits, the IRB should consider only those risks and benefits that may result from the research (as distinguished from risks and benefits of therapies subjects would receive even if not participating in the research). The IRB should not consider possible long-range effects of applying knowledge gained in the research (for example, the possible effects of the research on public policy) as among those research risks that fall within the purview of its responsibility.

3. Selection of subjects is equitable. In making this assessment the IRB should take into account the purposes of the research and the setting in which the research will be conducted and should be particularly cognizant of the special problems of research involving vulnerable populations, such as children, prisoners, pregnant women, mentally disabled persons, or economically or educationally disadvantaged persons.

4. Informed consent will be sought from each prospective subject or the subject's legally authorized representative, in accordance with, and to the extent required by 26.116.

5. Informed consent will be appropriately documented, in accordance with, and to the extent required by 26.117.

6. When appropriate, the research plan makes adequate provision for monitoring the data collected to ensure the safety of subjects.

7. When appropriate, there are adequate provisions to protect the privacy of subjects and to maintain the confidentiality of data.

8. When some or all of the subjects are likely to be vulnerable to coercion or undue influence, such as children, prisoners, pregnant women, mentally disabled persons, or economically or educationally disadvantaged persons, additional safeguards have been included in the study to protect the rights and welfare of these subjects (40 CFR 26.111).

Investigators and IRBs often struggle with the meaning of crucial terms, such as "minimal risk," "minor change," and "minor increase over minimal risk," on which key ethical and regulatory decisions rest (NBAC, 2001). Applying these regulatory requirements to nonclinical research (e.g., surveys) is even more difficult and cumbersome, because the limited regulatory detail provided is written in the context of clinical research (i.e., "that the research presents no more than minimal risk of harm to subjects and involves no procedures for which written consent is normally required outside of the research context" (40 CFR 117(c) (2)). As discussed in Chapter 4, the committee finds the concept of "minimal risk" to be of limited value as a guide to decision making in the context of the human dosing studies typically conducted for EPA regulatory purposes.

Balancing Risks and Probable Benefits

The principle of beneficence as elucidated in the *Belmont Report* states that persons should be "treated in an ethical manner not only by respecting their decisions and protecting them from harm, but also by making efforts to secure their well-being" (National Commission, 1979, 6). The principle requires that investigators attempt to maximize possible benefits and minimize possible harms. Federal regulations incorporate the obligation of beneficence by requiring IRBs to ensure that risks are minimized to the extent possible, given the research question, and are reasonable in relation to potential benefits to the participant or to the importance of the knowledge to be gained through the research (40 CFR 26.111(a)(1)-(2)).

Continual Review and Monitoring

Continual review and monitoring of research in progress is a critical part of the oversight system. Regular, continual review is necessary to ensure that emerging data or evidence have not altered the risk-benefit assessment so that risks are no longer reasonable. In addition, mechanisms should be in place to monitor adverse events, unanticipated problems, and changes to a protocol.

The regulations currently require that "an IRB shall conduct continuing review...at intervals appropriate to the degree of risk, but not less than once per year" (40 CFR 26.109(e)). However, the regulations do not specify the purpose or content of that review. In addition to the periodic reevaluation of risks and potential benefits as part of continuing review, IRBs conduct as-needed reviews when investigators request an amendment to approved protocols or in the event of unanticipated problems with a research study. Current regulations require institutions to create written procedures for "ensuring prompt reporting to the IRB of proposed changes in a research activity, and for ensuring that such changes in approved research, during the period for which IRB approval has already been given, may not be initiated without IRB review and approval except when necessary to eliminate apparent immediate hazards to the subject" (40 CFR 26.103(b) (4) (iii)). Institutions also are required to ensure that they report to the IRB "any unanticipated problems involving risks to subjects or...any suspension or termination of IRB approval" (40 CFR 26.103(b) (5)).

Other entities not considered in the federal Common Rule regulations, such as Data and Safety Monitoring Boards (DSMBs) or Data Monitoring Committees (DMCs), are beginning to play an increasingly important role in safety monitoring (DeMets et al., 1999; Fleming et al., 2002; FDA, 2001;

Gordon et al., 1998). These boards review data primarily from Phase 2 and 3 clinical trials from all participating sites and have access to unblinded data.[5]

Reporting Adverse Events

As mentioned previously, one of the requirements for approval of research is that IRBs must ensure that as ". . . appropriate, the research plan makes adequate provision for monitoring the data collected to ensure the safety of subjects" (40 CFR 26.111(a)(6)). FDA regulations are more specific than the Common Rule in delineating what must be reported and when. For FDA, all adverse events must be reported to sponsors during the three phases of product development, and serious unexpected adverse events must be reported by sponsors promptly to FDA and to all investigators. There are also mandatory postapproval reporting requirements. FDA may require sponsors to conduct Phase 4[6] (postapproval) studies to obtain further information about risks, potential benefits, and optimal use of a drug (21 CFR 312.85). Accumulating information on the public's experience with the approved drug or other FDA-regulated product can be reported to manufacturers, in which case it must be reported to FDA, or consumers may report their experiences directly to FDA (21 CFR 314.80, 314.81, 814.82, 814.34). FDA refers to this phase as postmarketing reporting.

[5]Phase 2 trials include controlled clinical studies conducted to evaluate a drug's effectiveness for a particular indication in patients with the disease or condition under study and to determine the common short-term side effects and risks associated with the drug. These studies are typically well controlled, closely monitored, and conducted with a relatively small number of patients, usually involving no more than several hundred subjects. Phase 3 trials involve the administration of a new drug to a larger number of patients in different clinical settings to determine its safety, effectiveness, and appropriate dosage. They are performed after preliminary evidence of effectiveness has been obtained and are intended to gather necessary additional information about effectiveness and safety for evaluating the overall risk-benefit relationship of the drug and to provide an adequate basis for physician labeling. In Phase 3 studies, the drug is used the way it would be administered when marketed. When these studies are completed and the sponsor believes that the drug is safe and effective under specific conditions, the sponsor applies to FDA for approval to market the drug. Phase 3 trials usually involve several hundred to several thousand patient-subjects.

[6]Concurrent with providing marketing approval, FDA may seek agreement from the sponsor to conduct certain postmarketing (Phase 4) studies to delineate additional information about a drug's risks, benefits, and optimal use. These studies could include, but would not be limited to, studying different doses or schedules of administration than were used in Phase 2 studies, use of the drug in other patient populations or other stages of the disease, or use of the drug over a longer period of time.

EPA also has statutory requirements for postmarket reporting by industry of adverse events resulting from the use of regulated chemicals or products (Federal Insecticide, Fungicide, and Rodenticide Act, §6(a) and the Toxic Substances Control Act, §8e).

MONITORING BY FEDERAL AGENCIES

Current mechanisms for monitoring include assurances of compliance issued by DHHS and several other federal departments, site inspections of IRBs conducted by FDA, other types of site inspections conducted by the funding agency, and institutional audits. Two primary federal agencies take the lead in monitoring human studies subject to the Common Rule: OHRP and FDA, both housed within DHHS.

Office for Human Research Protections

OHRP is charged with protecting research participants in biomedical and behavioral research conducted or sponsored by DHHS and other federal agencies that follow the Common Rule. The office operates on a system of Written Assurances of Compliance, in which the institution assures its compliance with the regulations as a condition of receiving federal research funds. If OHRP finds an institution to be noncompliant, it can suspend or revoke its assurance, stopping all or a portion of research activities at that institution.

Assurances are negotiated with each institutional grantee, with the negotiations allowing each institution to create its own policies and procedures for protection as long as they are fully consistent with federal regulations. The negotiation process also allows federal officials to educate institutions about requirements and procedures for participant protection.

The assurance indicates what an institution intends to do to protect research participants. In essence, it is a commitment on behalf of the institution to comply with all appropriate regulations and guidance in the conduct of all of its human research. Each federal department and agency may issue its own assurance, although many rely on DHHS assurances (NBAC, 2001). An assurance document is required for domestic institutions, and another assurance document is required for foreign institutions.

Food and Drug Administration

The most extensive system of data and safety monitoring exists in the area of clinical trials of drugs, medical devices, and other products subject to FDA review and approval. FDA inspects investigators, IRBs, and occasionally sponsors, to verify compliance with Good Clinical Practice (GCP) guidelines (FDA, 2003). FDA does not have the resources to inspect every investigator and thus is more likely to focus inspections on those entities that enroll large numbers of participants. Foreign investigators also are subject to inspection, but U.S. investigators are more likely to be scrutinized because of the logistics and available resources involved. Routine (not-for-cause) audits are essential elements of FDA's oversight. Research sponsors are expected to monitor the progress of studies, and investigators are required to maintain case histories for enrolled participants that include reports of serious adverse events. A distinct oversight unit within FDA provides ongoing surveillance of clinical research investigations. FDA's Bioresearch Monitoring Program audits the activities of clinical investigators, monitors, sponsors, and nonclinical (animal) laboratories. Its mission is to ensure the quality and integrity of data submitted to FDA for regulatory decisions, as well as to protect research participants.

The regulations that permit FDA to consider the protocols submitted to it during drug development are contained in 21 CFR 312 (human drugs) and 21 CFR 812 (medical devices). Federal regulations require that protocols submitted under an Investigational New Drug Application include detailed descriptions of the "clinical procedures, laboratory tests, or other measures to be taken to monitor the effects of the drug in human subjects and to minimize risk" (21 CFR 312.23). The submission of data, including the results of studies intended to support marketing, is required under 21 CFR 314. All relevant studies, such as drug studies that fail (i.e., that do not support the application or are incomplete), must be identified and submitted to FDA. FDA inspects study data to ensure their validity in support of an application, as well as the protection that was provided to the individuals from whom the data were collected. FDA may also audit the IRB of record for an inspected study, as well as investigate consumer complaints or reports from whistleblowers. If FDA finds that an investigator is noncompliant, he or she can be disqualified from future studies.

In the case of drugs and medical device trials, FDA inspections of clinical investigators generally are conducted after the trial is completed and a new drug application or premarket approval application for a medical device has been submitted for review, reflecting FDA's focus on assuring data quality.

In November 2001, FDA issued draft guidance entitled *Guidance for Clinical Trial Sponsors: On the Establishment and Operation of Clinical Trial Data Monitoring Committees*. According to FDA, the sponsor is responsible for ensuring that a DMC or DSMB (if applicable) operates under appropriate procedures. These boards are charged with reviewing interim data to determine whether the study should continue or be stopped for safety or therapeutic reasons according to pre-established stopping rules. The guidance document offers some perspective on criteria for establishing a DMC/DSMB, including committee composition, conflict of interest considerations, and other general considerations.

FDA also conducts surveillance (routine) and directed (when information "calls into question" regulated practices) inspections of IRBs. Usually IRB inspections are scheduled every five years, although if there are major problems, inspections can occur more frequently (FDA, 1998). During an inspection, an FDA field investigator (inspector) chooses a few studies that received initial IRB review within the past three years and follows them through the IRB review process. Inspectors look at IRB policies and procedures; minutes; membership; and records of studies, including protocol, consent form, investigator's brochure, and correspondence between the IRB and investigator. IRBs that are found to be out of compliance may be subjected to sanctions ranging from a warning letter to rejection of the data from the trial to prosecution (FDA, 1998).

The agency requires investigators to provide a written commitment that, before initiating an investigation subject to an institutional review requirement under 21 CFR 56, an IRB will review and approve the investigation in accordance with the regulations.

NONGOVERNMENTAL ACCREDITATION PROGRAMS

In recent years, there has been growing interest in nongovernmental performance-based accreditation systems to facilitate an emphasis on outcome measures in institutional research participants' protection programs and to meet evolving program needs. Participation in accreditation programs is a form of quality assurance, as efforts to prepare to meet accreditation standards should ordinarily have beneficial effects, and at a minimum, can help ensure that research programs conduct self-assessments, presumably noting and addressing deficient areas (IOM, 2001).

New accreditation organizations, such as the Association for the Accreditation of Human Research Protection Programs and the National Committee on Quality Assurance (NCQA), have appeared and are in the early phases of developing processes of setting and testing standards, with several institutions already having applied for accreditation status. In 2003 NCQA joined forces with the Joint Commission on Accreditation of Healthcare Organizations to form a new entity, the Partnership for Human Research Protection.

ENFORCEMENT

Each federal department that adheres to the Common Rule has the authority to enforce its own codification of the rule for research it conducts or sponsors. However, federal agencies and institutions with assurances of compliance from OHRP are subject to enforcement from that office as well. In the case of DHHS grantees and contractors, the enforcement authority is clear because OHRP is part of DHHS. But, when the assurance holder is the grantee of another department, OHRP decisions come from outside the regular reporting line of authority. Additionally, departments that use the OHRP assurance process may also have their own separate systems for enforcement, and there is little coordination among the various offices responsible for ensuring compliance with the Common Rule.

Federal regulations give department and agency heads the authority to terminate or suspend funding for research projects that are not in compliance with the regulations (40 CFR 26.123(a)). Common enforcement tools are the requirement of written responses or the enactment of specific changes to address the identified deficiencies; those who grant assurances also can restrict or suspend institutional assurances. Under its regulations, FDA, for example, can put new studies on hold (i.e., not permit them to proceed), prohibit enrollment of new participants, and terminate studies. FDA also can issue warning letters and restrict or disqualify investigators, IRBs, or institutions from conducting or reviewing research with investigational products.

RECENT CONCERNS ABOUT HUMAN RESEARCH PARTICIPANTS

Recent debate and analysis concerning the protection of research participants has focused on the federal and local institutions and agencies charged with this task, including federal regulatory agencies, academic and industrial laboratories, IRBs, and funding organizations. In particular, in the late 1990s examinations focused on IRBs. In June 1998, the Office of Inspector General (OIG) of DHHS issued a report, *Institutional Review Boards: A Time for Reform* (DHHS OIG, 1998), which stated that the effectiveness of IRBs is in jeopardy due to overwhelming demands. OIG concluded that the system, originally devised as a voluntary effort to oversee a much smaller research effort in the 1970s, was having difficulty contending with its growing and broadening workload with scant resources.

At the institutional level, OHRP increasingly imposed sanctions on institutions when it found systematic deficiencies or had concerns regarding systemic protections for research participants. The deficiencies concerned IRB membership; education of IRB members and investigators; institutional commitment; initial and continuing review of protocols by IRBs; review of protocols involving vulnerable persons; or procedures for obtaining voluntary informed consent. In 2001, NBAC issued a comprehensive report on ethical and policy issues in human research. The report recommended that federal oversight be centralized and that various components of the oversight system be revised to clarify regulatory responsibilities and to provide more guidance to assist institutions in formulating and implementing policies (2001).

In 2003 the Institute of Medicine (IOM) issued a report, *Responsible Research: A Systems Approach to Protecting Research Participants,* which provided an ethical and regulatory framework for institutions to create a system of protections involving investigators, research sponsors, research institutions, health care providers, federal agencies, and patient and consumer groups. The IOM report was in part written in response to systemwide concerns expressed by investigators, research institutions, IRBs, and others. Investigators and research institutions were complaining that there is a lack of national guidance on the administrative and ethical requirements of providing adequate protections and that the current federal posture is reactive and punitive rather than proactive and positive. Institutions were complaining about an overemphasis on documentation, which can lead to unproductive use of time that would be better spent seeking substantive protections. IRBs were complaining that the regulatory language is not easily understood and that federal regulators and research sponsors often interpret this language in ways that differ from local views. Because the IRB system operates at the local level, variation exists in how these boards operate and in the decisions they might make regarding a given protocol. Although this variation reflects the intent of the original regulations to insert local norms into the review process, some are concerned that this decentralization creates an untenable diversity of expectations for the approval process for multisite studies (IOM, 2003; NBAC, 2001).

IRBs themselves are overburdened and at times focus on avoiding risk in the face of rising regulatory pressures. IRB members, who must also fulfill other professional duties and who are often ill rewarded for their IRB service, are reviewing growing numbers of increasingly complex studies that may be conducted at multiple sites and reviewed by multiple IRBs (IOM, 2003).

The IOM committee also noted that research participants too often report that "they do not understand the nature or risks of research, that they find the informed consent process confusing, and that they are frequently divorced from the decision-making processes involved in the conduct of research" (2003, 39-40). It noted that informed consent documents have become increasingly complex and legalistic and too often are used inappropriately to protect the institution rather than the participant. The committee suggested that legal issues be separated from the consent process.

Finally, the IOM committee asserted that the scientific and ethical review of protocols should be equally rigorous. Because IRBs often are not equipped to assess the technical merits of a proposal and because scientific issues can become the focus of debate rather than ethical considerations, the committee recommended that a separate, distinctive review of the scientific merit of a protocol be conducted prior to review by an IRB.

OTHER ETHICAL FRAMEWORKS

Of note, other nonfederal, nonbinding guidelines for the protection of humans in research also are available, many of which were developed by the international community. In addition to the Nuremberg Code (1949), the *Declaration of Helsinki* (WMA, 2002) specifies requirements for voluntary participation of research participants, informed consent, and independent review of protocols. The declaration contains 32 statements of principle to guide medical research. Its conceptual framework is the medical ethics of the doctor-patient relationship, which is extended to research through the investigator-participant relationship. Other international guidelines, such as those provided by the International Conference on Harmonisation and the Council for International Organizations of Medical Sciences, provide detailed guidelines specific to drug trials and for GCP. The International Conference on Harmonisation was formed in 1990 and involves government drug regulation authorities and pharmaceutical trade organizations from the European Union, Japan, and the United States. Its guidelines have been adopted formally by FDA (ICH, 1996).

Thus, even though a particular study might not be subject to U.S. regulatory requirements, sponsors or investigators might voluntarily comply with the regulations or with the guidelines widely accepted in the international research community. Moreover, if the study is to be used to support marketing or investigational use in the United States, it must show compliance with ethical and scientific norms (21 CFR 312.120).

SUMMARY

The federal government regulates research involving humans through the Common Rule, which builds on the ethical principles articulated in international and national documents over the past 50 years. The regulations rest on two principal objectives in the oversight of human research: the conduct of independent review of research protocols by IRBs and the provision of voluntary informed consent to participate in research. The regulations are enforced by 16 agencies that conduct or sponsor human research.

The federal regulations provide a framework for considering risks and potential benefits, conducting review and monitoring activities, and reporting adverse events. They also specify the conditions under which informed consent must be obtained and the substantive requirements of consent. Monitoring of institutional activities is conducted at the federal level, and agencies employ various mechanisms for enforcement.

REFERENCES

Advisory Committee on Human Radiation Experiments (ACHRE). 1995. *Advisory Committee on Human Radiation Experiments—Final Report.* Washington, D.C.: U.S. Government Printing Office.

American Medical Association (AMA) Judicial Council. 1946. Supplementary Report of the Judicial Council of the American Medical Association. *Journal of the American Medical Association* 132:1090.

Beecher, H. K. 1966. Ethics and Clinical Research. *New England Journal of Medicine* 247(24): 1354-1360.

DeMets, D. L., S. J. Pocock, and D. G. Julian. 1999. The agonizing negative trend in monitoring clinical trials. *The Lancet* 354(9194):1983-1988.

Department of Health and Human Services. Office of Inspector General (DHHS OIG). 1998. *Institutional Review Boards: A Time for Reform.* Report No. OEI-01-97-00193. Washington, D.C.: DHHS.

Department of Health, Education, and Welfare (DHEW). 1971. *The Institutional Guide to DHEW Policy on Protection of Human Subjects.* Washington, D.C.: U.S. Government Printing Office.

DHEW. 1974. Protection of Human Subjects. *Federal Register* 39:18914-18920.

Faden, R. R., and T. L. Beauchamp. 1986. *A History and Theory of Informed Consent.* New York: Oxford University Press.

Fleming, T. R., S. Ellenberg., and D. L. DeMets, 2002. Monitoring clinical trials: issues and controversies regarding confidentiality. *Statistics in Medicine* 21(19):2843-2851.

Food and Drug Administration (FDA). 1998. *Guideline for the Monitoring of Clinical Investigators.* Available at www:fda.gov/ora/compliance_ref/bimo/clinguid.html.

FDA. 2001. Draft Guidance for Clinical Trial Sponsors: On the Establishment and Operation of Clinical Trial Data Monitoring Committees. Available at www.fda.gov/cber/gdlns/clindatmon.pdf.

Food and Drug Administration (FDA). 2003. *Good Clinical Practices.* Available at www.fda.gov/oc/gcp.

Gordon, V., J. Sugarman, and N. Kass. 1998. Toward a more comprehensive approach to protecting human subjects. *IRB: A Review of Human Subjects Research* 20(1):1-5.

Heller, J. July 26, 1972. Syphilis victims in U.S. study went untreated for 40 years. *New York Times,* Sec. A-1.

Institute of Medicine (IOM). 2001. *Preserving Public Trust: Accreditation and Human Research Participation* Protection *Programs.* Washington, D.C.: National Academy Press.

IOM. 2003. *Responsible Research: A Systems Approach to Protecting Research Participants.* Washington, D.C.: The National Academies Press.

International Conference on Harmonisation (ICH) of Technical Requirements for Registration of Pharmaceuticals for Human Use. 1996. *ICH Harmonized Tripartite Guideline. Guideline for Good Clinical Practice.* Geneva: ICH Secretariat, International Federation for Pharmaceutical Manufacturers Association.

Jones, J. H. 1981. *Bad Blood: The Tuskegee Syphilis Experiment.* New York: The Free Press.

Jonsen, A. R. 1998. *The Birth of Bioethics.* New York: Oxford University Press.

Moreno, J. D. 1999. *Undue Risk: Secret State Experiments on Humans.* New York: W. H. Freeman.

National Bioethics Advisory Commission (NBAC). 2001. *Ethical and Policy Issues in Research Involving Human Participants: Vol. 1.* Bethesda, MD: U.S. Government Printing Office.

National Commission for the Protection of Human Subjects of Biomedical and Behavioral Research (National Commission). 1979. *Belmont Report: Ethical Principles and Guidelines for the Protection of Human Subjects of Research.* Washington, D.C.: U.S. Government Printing Office.

National Institutes of Health (NIH). 1953. Group consideration of clinical research procedures deviating from accepted medical practice or involving unusual hazard. In: *Final Report, Supplemental Vol. 1,* 321-324. Washington, D.C.: U.S. Government Printing Office.

Nuremberg Code. 1949. *Trials of War Criminals before the Nuremberg Military Tribunals under Control Council Law No. 10, Vol. 2,* 181-182. Washington, D.C.: U.S. Government Printing Office.

President's Commission for the Study of Ethical Problems in Medical and Biomedical and Behavioral Research (President's Commission). 1981. *Protecting Human Subjects. First Biennial Report on the Adequacy and Uniformity of Federal Rules and Policies, and of Their Implementation for the Protection of Human Subjects.* Washington, D.C.: U.S. Government Printing Office.

President's Commission. 1983. *Implementing Human Research Regulations. Second Biennial Report on the Adequacy and Uniformity of Federal Rules and Policies, and of Their Implementation for the Protection of Human Subjects.* Washington, D.C.: U.S. Government Printing Office.

Public Health Service (PHS). 1966. Clinical investigations using human subjects. In: *Final Report, Supplemental Vol. 1,* 473-474. Washington, D.C.: U.S. Government Printing Office.

Stewart, W. H. 1966. Clinical research investigations using human subjects. In: *Final Report, Supplemental Vol. 1,* 473-474. Washington, D.C.: U.S. Government Printing Office.

Tuskegee Syphilis Study Ad Hoc Advisory Panel. 1973. *Final Report of the Tuskegee Syphilis Study Ad Hoc Advisory Panel.* Washington, D.C.: DHEW.

Wilson, C. 1953. Memorandum for the Secretary of the Army, Secretary of the Navy, Secretary of the Air Force. In *Final Report, Supplemental Vol. 1,* 308-310. Washington, D.C.: U.S. Government Printing Office.

World Medical Association (WMA). 2002. *Declaration of Helsinki: Ethical Principles for Medical Research Involving Human Subjects* (adopted 18th WMA General Assembly. Helsinki, Finland, June 1964; amended: 29th WMA General Assembly, Tokyo, Japan, October 1975; 35th WMA General Assembly, Venice, Italy, October 1983; 41st WMA General Assembly, Hong Kong, September 1989; 48th WMA General Assembly, Somerset West, Republic of South Africa, October 1996; and 52nd WMA General Assembly, Edinburgh, Scotland, October 2000. Note of Clarification on Paragraph 29 added by the WMA General Assembly, Washington 2002). Ferney-Voltaire, France: WMA.

3

Scientific Justification for and Conduct of Intentional Human Dosing Studies

INTRODUCTION

Scientific and ethical issues must be considered whenever intentional human dosing studies are proposed. These issues are, in most respects, interconnected. For example, an intentional human dosing study conducted for Environmental Protection Agency (EPA) regulatory purposes that is designed in such a way that it cannot make a scientifically sound contribution to regulatory decision making cannot be judged as ethical. However, for ease of explication, scientific and ethical issues are discussed separately in this chapter, with scientific issues the principal concern.

Intentional human dosing studies involving potentially toxic substances can, in some circumstances, contribute significant and useful knowledge for regulatory standard setting and other forms of public health protection. In fact, there is a long history of using data from such studies for these purposes, along with data from epidemiological investigations and animal experiments (Faustman and Omenn, 2001; Lippman et al., 2003; Paustenbach, 2002; Rodricks et al., 1997). The committee supports continued use of such information, provided that it is generated in compliance with the criteria and procedures recommended in this report that are designed to ensure ethical and scientific validity. The committee strongly recommends, however, that EPA should introduce much greater scientific care and rigor into its process for considering and relying on intentional human dosing studies by establishing criteria and procedures for deciding when and how they are to be conducted and their results used. Importantly, the same criteria and procedures should apply to both agency-conducted or agency-sponsored and third-party human dosing studies. Although EPA has in place procedures for ethical and patient protection review of agency-sponsored human studies (EPA, 1999), a more uniform and scientifically rigorous system should be considered for them and for third-party studies (discussed further in Chapter 6). The principal criteria for the scientific review of human dosing studies are briefly described in this chapter.

SCIENTIFIC ELEMENTS

As with all types of research, proposals to conduct intentional human dosing studies should begin with a discussion of the purpose and value of the study—the study justification. Assuming a study is justified, questions arise regarding study design and conduct and the reporting and evaluation of study results, matters that should be detailed in a study protocol. The protocol also includes information regarding protection of research participants. These two critical elements—study justification and study protocol—are the focus of this chapter.

It is important to recognize some of the critical distinctions between the types of research that are of interest to EPA as it carries out its legislative mandate and research that has a broader purpose. EPA is a regulatory agency that seeks information to fulfill its mission, such as that needed to improve the scientific basis of the risk assessments that are used to set regulatory standards or to fashion other types of health protection goals. Much of the committee's thinking regarding study justification and study protocols has been developed in recognition of the unique needs of regulatory agencies such as EPA. The committee also recognizes that all human research, whatever the purpose, must be conducted in adherence to the highest scientific standards, and it sought to incorporate such standards, along with those uniquely related to the regulatory process, into its recommendations. In addition, the committee proposes careful, independent review of study justifications and protocols for all intentional human dosing studies within the scope of EPA's mandate.

Before examining the issues involved in providing scientific justifications of and study protocols for intentional human dosing studies, a brief discussion is presented of the types of scientific investigations involving intentional dosing that are typically considered for possible conduct in human populations.

TYPES OF INTENTIONAL HUMAN DOSING STUDIES

There are three principal types of studies involving intentional dosing of research participants with chemicals that have been conducted for EPA regulatory purposes. The three types of studies seek to elicit (1) pharmacokinetic (PK) information, (2) effects on a biomarker, but not symptoms, and (3) effects on a symptom. These studies are not intended or expected to cause any irreversible or serious effect, based on previous animal and human experience. This is appropriate, as the committee cannot envision circumstances in which it would be ethical to knowingly harm research participants in order to generate data for EPA regulatory purposes. Although the three types of studies are not considered likely to cause lasting or serious harm to study participants, as will be explained in this chapter, their low levels of risk are not identical.

Studies That Seek to Elicit Pharmacokinetic Information

The goal of studies that seek to elicit PK information, or PK studies, is to delineate the absorption, distribution, metabolism, and excretion of chemicals in the body. Gaining an understanding of these processes can greatly aid in the interpretation of toxicity study findings and in the refinement of risk-assessment practices.

Comprehensive PK data can substantially reduce uncertainties inherent in route-to-route, high-to-low dose, and species-to-species extrapolations (Andersen, 1995; Leung and Paustenbach, 1995; see also Appendix B). In addition, knowledge of the toxicity and pharmacokinetics of a particular pesticide in one species can be useful in predicting and understanding adverse effects in a second species, whether in another laboratory animal or in humans.

A recent development in risk assessment is the use of so-called physiologically based PK models to improve the bases for cross-species extrapolation (Andersen, 2003; Bailer and Dankovic, 1997; Clewell et al., 2002). However, the successful development of such models depends on the availability of PK data in humans, with these and other developments in risk assessment placing increased reliance on human PK data. In addition to informing interspecies comparisons, human PK data can shed light on the appropriateness of the intraspecies uncertainty factor, for example, by showing similar PK activity in a wide range of participants.

Useful PK data typically can be developed in humans using very low doses of chemicals—doses that cannot cause adverse effects and often that cannot cause any detectable biological changes in research participants. PK studies conducted at levels that, based on extensive previous testing in animals, are expected to cause no detectable biological effect in participants, can be considered to pose no identifiable risks to research participants.[1]

[1]Phase 1 clinical trials for pharmaceuticals include PK studies, but these are undertaken at drug doses that are in the therapeutic range and at which some side effects may be observed. Such studies are distinguishable from those relevant to nonpharmaceutical substances.

Pharmacodynamic Studies That Examine Low-Dose Effects on Biomarkers

Pharmacodynamic (PD) studies (sometimes called toxicodynamic studies) are designed to measure the effect of a chemical or its metabolites on particular components of the body (e.g., tissues, cells, cell components). In some cases, the measured effect is a short-term biological response that is not thought to be adverse to health at the level studied, but that would cause the expected adverse effect of the chemical if the response were larger or more sustained—that is, it is on the causal pathway of the adverse effect. At the doses and duration used, however, the response would not be expected to cause an adverse effect in study participants. These biological responses often are referred to as biomarkers for the effects of interest (e.g., neurological effects).

Ordinarily, such studies—which involve brief and low exposure to chemicals and which are the majority of third-party studies submitted to EPA to date—present little risk to participants. Examples of such PD studies submitted to EPA include organophosphate (OP) pesticide inhibition of blood cholinesterase and perchlorate inhibition of radioactive iodine uptake by the thyroid gland (Greer et al., 2000; Lawrence et al., 2000; Lawrence et al., 2001). In each case, the inhibition is linked to the mechanism of the serious toxic effects of the chemical, but the effects on the biomarkers are known through other studies to become observable at dosage levels well below those at which adverse effects become clinically apparent. Moreover, these changes in biomarkers are reversible and temporary (whether longer term effects are possible is another consideration). The inhibition studies are valuable because the dose or blood concentration that causes a given degree of inhibition in humans and animals can be compared, which allows for the determination of different sensitivities to the inhibition among species.

In many of these studies, the specific determinations of interest in humans are the doses causing some effect on the biomarker (the lowest observed effect level, or LOEL) and the highest level at which no effect is seen (the NOEL, or no observed effect level) (NRC, 1994). These can then be compared with the LOEL and the NOEL in animals. Importantly, a study in which no effect is seen and no LOEL is defined is generally uninterpretable, because there is no evidence that the study could detect the effect on the biomarker and that the dose that was studied is truly the highest dose that causes no effect.

Pharmacodynamic Studies That Examine Low-Dose Effects That May Be Adverse to Participants

Some PD studies conducted for EPA regulatory purposes involve measuring the effect of an administered substance on a clinically detectable, adverse effect that, if larger and sustained could harm study participants. Such studies can yield a lowest observed adverse effect level (LOAEL), a no observed adverse effect level (NOAEL), and possibly an NOEL, although it is expected that any observed effects will not be sustained once exposure ceases (that is, the change is fully reversible). These studies present a somewhat greater risk to participants than PK studies, but, if the effects are well understood, familiar, closely observed, and reversible, participants should experience no lasting harm. The endpoints studied to date have involved air pollutants and have included changes in lung function or exercise ability and symptom onset (e.g., dyspnea) (Koenig et al., 1994; Langley et al., 2003). Generally, the substances studied are those to which the general population is already exposed, such as air and water contaminants. (See Box 3.1, which presents the committee's use of risk terminology for data derived from intentional dosing studies.)

In some studies, participants are healthy volunteers. In others, participants have a pre-existing medical condition (e.g., compromised cardiac or pulmonary function), and an exacerbation of the condition is used to assess exposure effects. If the purpose of such a study is to determine the effects of exposure on those who have pre-existing conditions that already put them at risk, it may be appropriate to include these participants. Additionally, in some cases valuable information can be gained from studies that include people with pre-existing conditions that are conducted at exposure levels known to exist in certain geographical areas, as participants would be exposed to levels they might encounter in their normal environments. Experimental studies of transmission dynamics that would include studies to determine infective dose, dose response curves, infectivity, and challenge studies (e.g., for cryptosporidium) are similar to those in this category of studies, as they are often expected to provoke a specific adverse but reversible effect (e.g., diarrhea).

All three of these types of studies can provide the opportunity to produce human data to improve the EPA risk-assessment process. In all cases, it is presumed that thorough animal data concerning the effects of the toxicants have been obtained and considered as the basis for concluding that a study poses no identifiable risk, that there is a reasonable certainty that no harm will occur to participants, or that the risks involved in the study are understood sufficiently that they can be evaluated in relation to the potential benefits. Of particular interest is whether the carcinogenicity and genotoxicity of a particular toxicant have been assessed. Depending

BOX 3.1
Committee's Use of Risk Terminology for Data Derived from Intentional Human Dosing Studies

No observed effect level ($NOEL_{HU}$)

A $NOEL_{HU}$ is the highest dose or concentration at which no changes of any kind are seen relative to controls. Depending on the number of doses studied and the ability to detect the $LOEL_{HU}$, the $NOEL_{HU}$ could underestimate the actual dose that could be given without a response.

Lowest observed effect level ($LOEL_{HU}$)/ No observed adverse effect level ($NOAEL_{HU}$)

A $LOEL_{HU}$ is the lowest dose or concentration at which a biological effect that is not adverse is seen. An example of such an effect would be cholinesterase inhibition by pesticides. A small amount of cholinesterase activity has not been demonstrated to have any adverse health effects. If lower doses are not studied, the $LOEL_{HU}$ could overestimate the dose that could actually elicit a response. What the committee terms a $LOEL_{HU}$ is often referred to by EPA as a no observed adverse effect level ($NOAEL_{HU}$). The committee is careful in its use of the term "$NOAEL_{HU}$" because it is most appropriately used in situations in which a clear $LOAEL_{HU}$ has been identified. A $NOAEL_{HU}$ is the highest dose or concentration at which no adverse effect is seen relative to controls.

Lowest observed adverse effect level ($LOAEL_{HU}$)

A $LOAEL_{HU}$ is the lowest dose or concentration at which an adverse effect is seen. In terms of the committee's discussion, for intentional human dosing studies there should be high confidence that any anticipated adverse effect *is not serious* and is *reversible.*

on the findings, evidence relating to genotoxicity or carcinogenicity in animals could be important in determining whether human studies are safe enough to conduct and could influence the content of the informed consent.

It is important to underscore the difference between the three types of studies described here and clinical trials involving therapeutic doses of experimental drugs. It is well recognized and accepted that even Food and Drug Administration (FDA)-approved drugs can pose significant risks to patients and thus that, in Phase 2 and Phase 3 clinical trials on experimental drugs, research participants may experience adverse side effects. Indeed, in addition to assessing a drug's effectiveness, these trials are used to identify and better understand its possible harmful side effects. This possibility of harm is one reason informed consent and independent Institutional Review Board (IRB) review of the risks and benefits of a trial are needed.

Importantly, in therapeutic clinical trials, there may be personal benefits for study participants, sometimes as an immediate consequence of participation, more typically in developing treatments for the condition the participant has. This benefit can, in some cases, be considered in deciding whether the risks are justified. The human dosing studies likely to be conducted (and found ethically acceptable) for EPA regulatory purposes pose much less risk to participants than often is accepted in drug trials, but they also are unlikely to provide any personal benefit to the participants. The different character of both the risks and benefits in human dosing studies conducted for EPA regulatory purposes makes many of the specific issues addressed in this report novel and underlies many of the committee's recommendations.

This chapter now turns to the issues of how and why such studies may be justified and the types of protocols that are needed to ensure their proper conduct, including the protection of research participants.

JUSTIFICATION FOR INTENTIONAL HUMAN DOSING STUDIES

Criteria for Study Justification

Justification of intentional human dosing studies depends on the importance of the expected results to a regulatory decision that will protect the public health and a demonstration that other means of acquiring the necessary information are substantially deficient. In the case of intentional human dosing studies conducted for EPA regulatory purposes, ethical and scientific standards demand that every effort be made in advance to ensure that the biological endpoints to be measured are important to the assessment of human risk. Whether the data are to be used for determining risks for acute or short-term exposures, or for the derivation of a Reference Concentration (RfC) or a Reference Dose (RfD), every effort should be made to document in advance their critical nature. Data unrelated to or peripheral to regulatory risk assessments should never be sought through intentional human dosing studies, even those involving no identifiable risk to participants (PK studies).

For example, cholinesterase inhibition is generally considered to be the mechanism of action of the neurotoxic effects of many organophosphates (OP) pesticides, and doses that do not inhibit acetylcholinesterase (AChE) do not produce the cholinergic-mediated effects (see IOM, 2003, for review). The inhibition of cholinesterase that mediates toxicity occurs at the synapses of the central and peripheral nerves, but in human studies only blood cholinesterase activity usually is measured. It is necessary, therefore, to know, in considering whether such a study is justified, if blood cholinesterase is a relevant measure of the state of peripheral nerve and central cholinesterase.

Even if acute blood cholinesterase inhibition were considered a reasonable surrogate marker for acute toxicity, it might not be an adequate marker for all effects of OPs, including possible long-term effects or effects on development. In addition, effects might differ across age groups or developmental stages (Clewell et al., 2002). This issue is sometimes far from straightforward.[2]

---

[2]For example, acetycholinesterase (AChE) inhibition is considered to be the primary mechanism of the acute neurotoxicity of OP pesticides, although some OPs have additional modes of action (Mileson et al., 1998). Doses of OPs that produce modest decrements in AChE activity are generally accepted to be substantially lower than doses required to elicit clinically recognizable cholinergic-mediated effects. Nonetheless, inhibition thresholds for the onset of particular effects are often controversial. AChE inhibition that mediates neurotoxicity occurs at synapses of central and peripheral nerves, but alterations in plasma and red blood cell cholinesterase activities are commonly monitored as indices of potential central effects in adults and children (Wessels et al., 2003). The relative sensitivities of the enzymes in plasma and in erythrocytes to an OP have been shown to be species dependent (Karanth and Pope, 2003). The relevance of such findings to man is subject to question, since human brain samples cannot be analyzed. An additional area of uncertainty is the relationship between AChE inhibition by OPs and chronic neurological effects (Committee on Toxicity of Chemicals in Food, Consumer Products, and the Environment, 1999; Steenland et al., 2000). It is beyond the purview of this committee, however, to evaluate and to make EPA policy recommendations on the state of science in these areas.

The age dependence of susceptibility to acute OP poisoning has received considerable attention. A number of research groups have observed that newborn rodents are the most susceptible to OP-induced AChE inhibition and accompanying signs of excessive cholinergic effects. In an early study, Benke and Murphy (1975) observed a progressive decrease in susceptibility to acute poisoning and an increase in metabolic detoxification of parathion and methyl parathion, with increasing age of rats from 1 to 63 days of age. Moser et al. (1998) demonstrated that activities of plasma and liver carboxylesterases and A-esterases, key detoxifying enzymes, are inversely proportional to chlorpyrifos-induced AChE inhibition and acute toxicity in maturing rats.

Researchers have recently examined effects of neonatal and juvenile OP exposure on a variety of neurochemical and behavioral parameters. Liu et al. (1999) found that methyl parathion, but not chlorpyrifos, produced more pronounced reductions in brain AChE activity and muscarinic binding in 7- than in 90-day-old rats. Some investigators (e.g., Levin et al., 2001) have reported that neonatal exposures to OPs induce long-term cognitive deficits in rats, although dose-response data are lacking. Other researchers (e.g., Maurissen et al., 2000) have seen no residual effects on learning or memory. It has been proposed that AChE promotes neural growth and differentiation, so that AChE inhibition by pesticides may disrupt cell replication, communication, and adhesion (Brimijoin and Koenigsberger, 1999).

Therefore, even if it were well established that the short- and long-term effects of OPs are mediated through cholinesterase inhibition and that a dose with no effect on blood cholinesterase is very unlikely to cause harm in adults, those data would not necessarily provide information regarding possible effects on the developing nervous system. Because of these issues, careful documentation of the value and relevance of the end-point to be measured is a critical component of study justification.

A second criterion that should be applied to justify studies in humans pertains to the availability of different ways of acquiring the necessary data. Data from animal models have widespread use in regulatory risk assessment, and many, if not most, standards are derived from such assessments (NRC, 1994). Considerable effort over the past several decades has been devoted to improving and standardizing protocols for animal bioassays (Ashby, 2001; Gaylor, 1996).

It may be asked why human studies are ever justifiable if animal models are available. There are three broad reasons to turn to human studies to supplement animal data (see Appendix A for further discussion of the limitations of animal studies). First, it is well established that animal models are not especially accurate predictors of certain adverse biological effects, particularly those involving immune-mediated responses (e.g., hypersensitivity reactions, other allergic responses) and certain airway responses to hazardous air pollutants (Samet et al., 1994). In some cases, no validated animal models may be available to serve as surrogates for individuals with compromised immune systems or with other medical conditions that may render them especially sensitive to pollutants. Many of the EPA-sponsored short-term air pollutant studies in humans have been motivated by such concerns, and the data derived from some of them have been informative for both setting standards and for gaining critical knowledge about mechanisms of toxicity (EPA, 2003).

Second, animal models have little value for assessing adverse effects that cannot be objectively measured, such as those that can be known only because they can be reported by study participants (headaches are a prime example, as are feelings of nausea and dizziness). Such symptoms can be significant indicators of toxicity, and sometimes efforts must be made to determine whether they can be produced by certain chemicals.

There have been reports that repeated OP exposures, that do not cause inhibition of brain AChE in preweanling rats, result in decreased locomotor activity and impaired spatial learning when the rats become juveniles (Carr et al., 2001; Jett et al., 2001). Efforts are now being made to understand the functional significance of cellular and molecular changes observed in the immature central nervous system and to determine whether household or dietary exposures to pesticides can produce such changes.

A third reason that animal data may be insufficient is that there are, in some instances, quantitative differences in response between average human and animal responses. This is recognized in the 10-fold interspecies uncertainty factor typically applied when animal data are used to set exposure limits (this assumes that humans may be 10 times more sensitive) (see Chapters 2 and 7), but, in fact, human sensitivity may be either greater than or less than 10 times that of animals. Human studies of a relevant endpoint can allow for decisions that are more informed about the risk of any given level of exposure (Dourson et al., 2001).

Even PK studies involving no identifiable risk to participants require scientific justification. As noted, PK data can be relevant to interspecies comparison and to within-human variability. The specific use and value of PK information need to be considered.

Documentation of Study Justification

Written and well-referenced documentation of the justification for intentional human dosing studies is a necessary prerequisite for their conduct. As will be seen in Chapter 6, the committee recommends that, prior to the conduct of both agency-sponsored and third-party studies, EPA should establish an independent board to review such documentation and to review the study protocols (see Box 3.2).

It should be emphasized that although a study may be scientifically justifiable according to the above criteria, it may nonetheless not be undertaken if the protection of research participants cannot be ensured. The committee views ensuring the protection of research participants as an element of the study protocol.

BOX 3.2
Criteria for Judging the Adequacy of the Scientific Justification for Intentional Human Dosing Studies

1. The proposed study is designed to yield data of direct relevance and importance to a risk-assessment process that is part of a regulatory standard-setting activity or another form of public health protection.
2. Alternative methods of obtaining the data are not available.
3. There is thorough, scientific documentation available to support criteria (1) and (2), and the documentation has undergone independent review by an IRB and the EPA Human Studies Review Board that is recommended in Chapter 6.

PROTOCOLS FOR INTENTIONAL HUMAN DOSING STUDIES

Along with providing documentation related to the justification of an intentional human dosing study, a study protocol should be provided that sets forth the study's design and method of conduct and a plan for analyzing, reporting, and evaluating the results. These elements must be described and justified.[3] The protocol also should include a demonstration of how participant protection will be assured.

There is an extensive literature on the design, conduct, and analysis of clinical studies (see, for example, FDA, 2003). However, rather than provide a comprehensive treatment here, the committee highlights issues that are especially important to the evaluation of intentional human dosing studies or that were identified as especially problematic in studies that were submitted to EPA and reviewed by the committee.

Overall Study Plan

The specific objectives of the proposed study, as described in the scientific justification document, are used to guide study design. Selection of doses to be used, criteria for participant selection, sizes of individual groups, and clinical measurements to be made are all dictated by the stated objectives of the study. In the end, it must be shown that the proposed study design is capable of yielding results that will satisfy the specified objectives.

A plan for the specific procedures to be followed in the conduct of the study, and for recording all of the relevant data, also is necessary, as is a description of methods to be used in evaluating study results. Finally, the overall plan should include documentation of the adequacy of preclinical data for establishing that study participants are not likely to be harmed at the doses selected and that other appropriate safeguards are in place.

Aspects of Study Design

Five features are critical to designing an intentional human dosing study, including endpoint, dose, and participant selection; study method; and dosing and measurement schedules.

---

[3]The committee notes that for most of the third-party studies on cholinesterase inhibition received by the EPA, protocols did not contain scientific support for many of the study designs and methods selected.

1. *Endpoint Selection*

The endpoints to be measured should be described and their relation to study objectives explained. It should be asked whether the endpoints are the same as, or human equivalents to, those assessed in animals. The ability to measure the selected endpoints with reliability and precision should be described.

2. *Dose Selection*

Sufficient preclinical (animal) data relevant to the clinical endpoint of interest, or other human data, should be available to support selection of the doses to be used in humans. Dose selection for PK studies usually is dictated by technical questions related to analytical detection capabilities, rather than by any factors related to clinical response. The highest dose selected should be sufficient to induce the desired response, whether it is a critical biomarker or other endpoint. Doses lower than the highest dose should be selected to characterize the dose-response relationship and, if possible, to identify the maximum dose that represents the NOEL. Failure to see any response raises the question of whether the study was able to detect the response at all—that is, did the study have assay sensitivity? Consideration also must be given to the purity of the test compound, to ensure that it differs in no significant way from that of the test compound used in the preclinical studies that were used as the basis for dose selection. The mode of compound administration also should be described and the relevance of the method of administration justified. Box 3.3 provides two examples of designs used in studies submitted to EPA to identify a $NOEL_{HU}$, with accompanying committee commentary.

3. *Participant Selection*

The choice of participants is dictated by the objectives of the study. If the objective is to modify uncertainty factors and replace animal data with relevant human data (potentially eliminating the need for the uncertainty factor for animal-to-human extrapolation), healthy adult humans of, for example, similar age and weights might be most appropriate to represent the average human population. Selection of such individuals also would reduce possible variability in biological responses and make more precise estimates of the intraspecies factor possible. Although this study will not capture the full range of human variability, risk-assessment procedures already include an intraspecies uncertainty factor that will accommodate expected variability (see Chapters 2 and 7). Despite the desirability of a reasonably homogeneous population, including participants of both gen-

BOX 3.3
Examples of Study Designs That Identify a $NOEL_{HU}$

Design #1

The investigator selects 10 males: Caucasian, healthy, 20 to 30 years of age. Each participant is dosed with Xmg of the substance, and the relevant endpoint is measured. No effects of interest are seen.

Commentary on Design #1: If no effects are observed, it is not possible to use the Xmg dose as establishing the $NOEL_{HU}$, because in the absence of observed effects, there is no evidence that the study could detect an effect if it were present (no proof of assay sensitivity) and no information about the dose that did have an effect (the $LOEL_{HU}$ or $LOAEL_{HU}$). Furthermore, there could be no estimate of the uncertainties that surround any conclusion. This kind of study has been called a "NOEL-only" study and is not useful for formal risk assessment (see Chapter 7).

Design #2

After careful review of the preclinical data, the investigator expects that the $NOEL_{HU}$ for the substance will be near Xmg. Rather than dosing every participant with that dose, dosing proceeds as follows:

a) Six doses at and surrounding Xmg are selected for the study (0, X/a, X/b, X, bX, aX), where a>b>1 are appropriate factors suggested by the preclinical data. The 0mg dose creates a control group.

b) A random sample of individuals is recruited for the study, and participants are randomly allocated to the six dose groups.

Assume that at doses X/a and X/b, there are no observed effects different from the control, but that at dose X an effect is observed in some individuals. At the higher doses (bX and aX), the proportion of individuals who exhibit the effect is higher still.

Commentary on Design #2: The study produces information that supports estimating a dose-response curve and confidence limits for the curve. The analysis can reasonably and credibly establish the $NOEL_{HU}$, or $NOAEL_{HU}$, depending on the outcome measured for acute exposure.

ders is desirable, unless there is compelling evidence that differences in response are not expected.

If, on the other hand, the goal of the study is to set acceptable levels of an air pollutant, it will be critical to focus on sensitive populations because they represent those most clearly at risk and often include individuals with specific medical conditions. Careful review of these conditions among potential participants is critical in order to avoid wide variability among members of the study and control groups and to avoid including participants who will not test the question at issue (see Chapter 5).

Study protocols should include justification for participant selection, a description of how potential participants are identified, and a description of the procedures to be used in randomizing participants to dose groups.

4. *Study Method*

Protocols must provide a carefully delineated justification for the proposed study method. Sample sizes proposed for each group should be justified by a demonstration that there is adequate power to detect a relevant change in the endpoint(s) to be measured given the estimated variability in the response.

5. *Dosing and Measurement Schedules*

The specific schedule for dosing and measuring the response should be clearly related to the objectives of the study. Scientific support for the schedules should be provided.

Conducting and Recording Statistical Analysis of Results

It is essential to develop the statistical analysis plan as an integrated part of the study design and to ensure that primary statistical analysis is linked to primary study goals. An approach for recording the results also should be provided. All data generated should be thoroughly analyzed and reported, and the protocol should identify a hierarchy of outcomes with a narrowly defined set of primary goals. Confidence intervals surrounding the estimates and other measures of uncertainty should be reported, and the quality of the data should be assessed as they come in so that timely corrections can be made (and documented).

Protection of Research Participants

One section of the protocol should be devoted to a careful and thoroughly documented presentation of the likely risks to participants at the proposed levels and duration of dosing. This documentation should be accompanied by a discussion of other critical elements of study participant protection, as called for in Chapter 5 and 6.

The Protocol Document

As noted, the study protocol should provide a detailed statement on the study objectives and scientific justification for the study design. It also should provide the study analysis or a detailed statement on it and information regarding how the data will be reported. It should contain a thorough guide to participant protections, including any proposed data and safety monitoring plan or committee, assurance that the proposed study will be conducted in compliance with Good Clinical Practice (GCPs) guidelines, and provisions to permit EPA to monitor the conduct of the study (FDA, 2003). It also should include assurance of review by an IRB or an equivalent body as well as assurance that informed consent was obtained. Finally, the protocol should contain a copy of the written consent form, describe the consent procedures, and include an agreement to permit onsite inspection. The committee recommends that an independent review board evaluate the study protocol document, together with the scientific justification (see Chapter 6).

Study Reporting to EPA

For EPA to assess the scientific validity of the results of an intentional human dosing study, study reporting should be comprehensive and should include an assessment of the implications of the study relative to the study objectives and the relationship of study results to existing knowledge. The full protocol and detailed analyses should be submitted to EPA with a narrative interpretation of the results that includes summary tables and graphs, the data codebook, and all data (in a computer analyzable form, e.g., an SAS dataset), so that a reviewer could replicate reported analyses and conduct additional analyses. The report should fully document any problems and any changes in the protocol. Study participant characteristics must be well documented, and all adverse events must be reported and evaluated, regardless of determination of "relatedness" or causality assessment. All relevant studies conducted by the laboratory, clinic, or funding organization should be reported in at least summary form, even if their findings are not in the interest of those sponsoring the study.

Recommendation 3-1: Scientific Validity of Intentional Human Dosing Studies

EPA should issue guidelines for determining whether intentional human dosing studies have been:

a. justified, in advance of being conducted, as needed and as scientifically appropriate, in that they could contribute to address-

ing an important scientific or policy question that cannot be resolved on the basis of animal data or human observational data;

b. designed in accordance with current scientific standards and practices to (i) address the research question, (ii) include representative study populations for the endpoint in question, and (iii) meet requirements for adequate statistical power;

c. conducted in accordance with recognized good clinical practices, including appropriate monitoring for safety; and

d. reported comprehensively to EPA, including the full study protocol, all data produced in the study (including adverse events), and detailed analyses of the data.

SUMMARY

Three principal types of studies involving intentional dosing of research participants with chemicals have been conducted for EPA regulatory purposes: PK studies; PD studies of low-dose, nonadverse effects; and PD studies designed to elicit an adverse but fully reversible effect. The first two types of studies are likely to pose no identifiable risk to study participants or can be scientifically demonstrated to provide a reasonable certainty that no harm will result to them. PD studies eliciting adverse but reversible effects pose a risk, although it should remain low, depending on factors such as the nature of the effect and whether it is fully reversible, whether the study is properly conducted, and the study population.

Prior to its conduct, a study should be deemed justifiable on the basis of existing scientific data from animal and other studies. This justification should include an explanation of the relevance and importance of the endpoint to the potential effects of concern for regulatory purposes and evidence of the lack of ability to obtain the needed information in other ways.

An intentional human dosing study cannot be ethical if it is not designed, conducted, and reported in ways that ensure the highest scientific quality. The need for scientific quality begins at the planning stage and includes the choice of endpoint, exposure conditions, and dose, as well as a consideration of the study power and statistical analysis. The full study results should be reported, including details regarding design, conduct, and outcomes, even if they are not in the interest of those sponsoring the study.

REFERENCES

Andersen, M. E. 1995. Development of physiologically based pharmacokinetic and pharmacodynamic models for applications in toxicology and risk assessment. *Toxicology Letters* 79:35-44.

Andersen, M. E. 2003. Toxicokinetic modeling and its applications in chemical risk assessment. *Toxicology Letters* 138:9-27.

Ashby, J. 2001. Expectations for transgenic rodent cancer bioassay models. *Toxicologic Pathology* 29(Suppl.):177-182.

Bailer, A. J., and D. A. Dankovic. 1997. An introduction to the use of physiologically based pharmacokinetic models in risk assessment. *Statistical Methods in Medical Research* 6:341-358.

Benke, G. M., and S. D. Murphy. 1975. The influence of age on the toxicity and metabolism of methyl parathion and parathion in male and female rats. *Toxicology and Applied Pharmacology* 31:254-269.

Brimijoin, S., and C. Koenigsberger. 1999. Cholinesterases in neural development: new findings and toxicologic implications. *Environmental Health Perspectives* 107(Suppl. 1):59-64.

Carr, R. L., H. W. Chambers, J. A. Guarisco, J. R. Richardson, J. Tang, and J. E. Chambers. 2001. Effects of repeated oral postnatal exposure to chlorpyrifos on open-field behavior in juvenile rats. *Journal of Toxicological Sciences* 59:260-267.

Clewell, H. J., J. Teeguarden, T. McDonald, R. Sarangapani, J. E. Lawrence, T. Covington, R. Gentry, and A. Shipp. 2002. Review and evaluation of the potential impact of age- and specific pharmacokinetic differences on tissue dosimetry. *Critical Reviews in Toxicology* 32:329-389.

Committee on Toxicity of Chemicals in Food, Consumer Products, and the Environment. 1999. *Organophosphates: Report of the Working Group on Organophosphates*. Available at www.doh.gov.uk/cot/op.htm.

Dourson, M .L., M. E. Andersen, L. Erdreich, and J. Macgregor. 2001. Using human data to protect the public's health. *Regulatory Toxicology and Pharmacology* 33:234-256.

Environmental Protection Agency (EPA). 1999. *Policy and Procedures on Protection of Human Research Subjects in EPA Conducted or Supported Research*. EPA Order. Available at www.epa.gov/oamrtpnc/forms/1000_17a.pdf.

EPA. Risk Assessment Forum. 2003. *Draft Formal Guidelines for Carcinogen Risk Assessment*. EPA/630/P-03/001A, NCEA-F-0644A.

Faustman, E. M., and G. S. Omenn. 2001. Risk assessment. In: C. D. Klaassen, ed. *Casarett and Doull's Toxicology: The Basic Science of Poisons*. 6th ed. New York: McGraw Hill.

Food and Drug Administration (FDA). 2003. *Good Clinical Practices*. Available at www.fda.gov/oc/gcp.

Gaylor, D. W. 1996. Risk estimation—an overview: disease risk estimation based on animal bioassays. *Drug Metabolism Reviews* 28:9-27.

Greer, M. A., G. Goodman, R. Pleus, and S. E. Greer. 2000. Does environmental perchlorate exposure alter human thyroid function? determination of the dose-response for inhibition of radioiodine uptake. In: Abstracts of the 12th International Thyroid Congress, October; Kyoto, Japan. *Endocrine Journal* 47(suppl.):146.

Institute of Medicine (IOM). 2003. *Responsible Research: A Systems Approach to Protecting Research Participants*. Washington, D.C.: The National Academies Press.

Jett, D. A., R. V. Navoa, R. A. Beckles, and G. L. McLemore. 2001. Cognitive function and cholinergic neurochemistry in weanling rats exposed to chlorpyrifos. *Toxicology and Applied Pharmacology* 174:89-98.

Karanth, S., and C. Pope. 2003. In vitro inhibition of blood cholinesterase activities from horse, cow and rat by tetrachlorvinphos. *International Journal of Toxicology* 22:429-433.

Koenig, J. Q., D. S. Covert, W. E. Pierson, Q. S. Hanley, V. Rebolledo, K. Dumler, and S. E. McKinney. 1994. Oxidant and acid aerosol exposure in healthy subjects and subjects with asthma—part I: effects of oxidants, combined with sulfuric or nitric acid, on the pulmonary function of adolescents with asthma. *Research Report (Health Effects Institute)* 70:1-36.

Langley, S. J., S. Goldthorpe, M. Craven, J. Morris, A. Woodcock, and A. Custovic. 2003. Exposure and sensitization to indoor allergens: association with lung function, bronchial reactivity, and exhaled nitric oxide measures in asthma. *Journal of Allergy and Clinical Immunology* 112(2):362-368.

Lawrence, J. E., S. H. Lamm, and L. E. Braverman. 2001. Low dose perchlorate (3mg daily) and thyroid function [Letter]. *Thyroid* 11:295.

Lawrence, J. E., S. H. Lamm, S. Pino, K. Richman, and L.E. Braverman. 2000. The effect of short-term low-dose perchlorate on various aspects of thyroid function. *Thyroid* 10:659-663.

Leung, H. W., and D .J. Paustenbach. 1995. Physiologically based pharmacokinetic and pharmacodynamic modeling in health risk assessment and characterization of hazardous substances. *Toxicology Letter* 79:55-65.

Levin, E. D., N. Addy, A. Nakajima, N. C. Christopher, F. J. Seidler, and T. A. Slotkin. 2001. Persistent behavioral consequences of neonatal chlorpyrifos exposure in rats. *Developmental Brain Research* 130:83-89.

Lippman, M., B. S. Cohen, and R. B. Schlesinger. 2003. Risk assessment. In: *Environmental Health Science: Recognition, Evaluation, and Control of Chemical and Physical Health Hazards*. New York: Oxford.

Liu, J., K. Olivier, and C. N. Pope. 1999. Comparative neurochemical effects of repeated methyl parathion or chlorpyrifos exposures in neonatal and adult rats. *Toxicology and Applied Pharmacology* 158:186-196.

Maurissen, J. P. J., A. M. Hoberman, R. H. Garman, and T. R. Hanley, Jr. 2000. Lack of developmental neurotoxicity in rat pups from dams treated by gavage with chlorpyrifos. *Journal of Toxicological Sciences* 57:250-264.

Mileson, B. E., J. E. Chambers, W. L. Chen, W. Dettbarn, M. Ehrich, A. T. Eldefrawi, D. W. Gaylor, K. Hamernik, E. Hodgson, A. G. Karczmar, S. Padilla, C. N. Pope, R. J. Richardson, D. R. Saunders, L. P. Sheets, L. G. Sultatos, and K. B. Wallace. 1998. Common mechanism of toxicity: a case study of organophosphorus pesticides. *Journal of Toxicological Sciences* 41:8-20.

Moser, V. C., S. M. Chanda, S. R. Mortensen, and S. Padilla. 1998. Age- and gender-related differences in sensitivity to chlorpyrifos in the rat reflect developmental profiles of esterase activities. *Journal of Toxicological Sciences* 46:211-222.

National Research Council (NRC). 1994. *Science and Judgment in Risk Assessment*. Washington, D.C.: National Academy Press.

Paustenbach, D. 2002. Section A. Theory: human health assessment. In: *Human and Ecological Risk Assessment*. New York: John Wiley & Sons.

Rodricks, J. V., L. Rudenko, T. B. Starr, and Turnbull, D. 1997. Risk assessment. In: Sipes, I., C. McQueen, and A. Gandolfi, eds. *Comprehensive Toxicology* 315-338. Pergamon, Elsevier Sciences: Oxford, UK.

Samet, J. M., W. E. Pepelko, B. Sonawane, G. E. Hatch, K. E. Driscoll, and G. Oberdörster. 1994. Risk assessment of oxidant gases and particulate air pollution: uncertainties and research needs. *Environmental Health Perspectives* 10(Suppl.):209-213.

Steenland, K., R. B. Dick, R. J. Howell, D. W. Chrislip, C. J. Hines, T. M. Reid, E. Lehman, P. Laber, E. F. Krieg Jr., and C. Knott. 2000. Neurologic function among termiticide applicators exposed to cholorpyrifos. *Environmental Health Perspectives* 108:293-300.

Wessels, D., D. B. Barr, and B. Mendola. 2003. Use of biomarkers to indicate exposure of children to organophosphate pesticides: implications for a longitudinal study of children's environmental health. *Environmental Health Perspectives* 111:1939-1946.

4
A Risk-Benefit Framework for Assessing Intentional Human Dosing Studies

INTRODUCTION

As discussed in Chapter 2, the regulatory framework for human research imposes a number of fundamental conditions: (1) exposure of participants to any risk must be scientifically necessary; (2) risks to participants must be minimized; (3) the potential benefits from the research must justify any risks participants may face; (4) selection of participants must be equitable; (5) participants must give informed consent; and (6) an independent board must give prior approval to the research design and monitor compliance with procedures to protect participants. Research results submitted to the Environmental Protection Agency (EPA) must satisfy these conditions as a minimum condition for acceptability. This chapter examines the risks and benefits of intentional human dosing studies and considers when the benefits may justify the risks.

Comparing risks and benefits in human experiments is a critical and often difficult task. The National Bioethics Advisory Commission (NBAC) observed that "there are no clear criteria for IRBs to use in judging whether the risks of research are reasonable in relation to what might be gained by the research participants" (NBAC, 2001, 69). The task is particularly difficult in the case of human studies submitted to EPA for regulatory purposes, because the benefits of the research typically accrue not to the study participants, but to society at large, calling for an especially cautious approach in applying general principles. The committee decided that it could a provide a framework for clarifying some specific issues regarding the use of intentional human dosing studies for EPA regulatory decision-making purposes, but it made no pretense of being able to resolve all of the nettlesome issues, especially the potentially wide range of study-specific risk-benefit comparisons that might be raised in this context. These ultimately must be resolved through publicly transparent policy deliberations and through the case-by-case decisions made by duly constituted review bodies.

POTENTIAL BENEFITS FROM INTENTIONAL HUMAN DOSING STUDIES

The Common Rule under which EPA conducts and sponsors studies requires that "risks to subjects" must be "reasonable in relation to anticipated benefits, if any, to subjects, and the importance of the knowledge that may reasonably be expected to result" (40 CFR §26.111(a)(2)). NBAC interpreted the basic ethical framework guiding human research as requiring independent review to "ensure that risks are reasonable in relation to potential personal and societal benefits" (NBAC, 2001, 3).

As indicated by these formulations of the risk-benefit requirement, potential or anticipated benefits from studies involving humans can be divided into two broad types—personal and societal. Potential personal benefits are those that may accrue to an individual by virtue of participating in the experiment. Potential societal benefits are those that accrue to the society as a whole or to groups within a society by virtue of the application of the scientific results of the study.

For example, placebo-controlled Phase 3 drug trials are designed to test the effectiveness of a drug. If the drug proves effective, at least some of the participants have the prospect of receiving direct medical benefit from the new treatment. Both intervention and control participants also may have the prospect of gaining other personal benefits, although such benefits would not result from receiving the drug being studied. For example, participants may benefit from increased knowledge about their condition from the medical evaluation that is included in the study.

There are many clinical trials, however, that are not intended to offer direct clinical benefits to participants. Phase 1 drug trials, for example, are designed to test for side effects of a drug and to establish dosing regimens. These trials often enroll healthy individuals who do not suffer from the condition the drug is intended to treat. These participants will receive no direct medical benefits from receiving the drugs during the trial. Nonetheless, carefully designed and conducted Phase 1 trials with healthy volunteers have been considered ethically acceptable. When risks are minimized, some risks to informed and consenting participants can be and are considered reasonable in light of the potential societal benefits that may result from the study.

Personal Benefits

Experiments involving intentional human dosing that are conducted for EPA's regulatory purposes do not present the possibility of providing any health-related direct personal benefits to participants. As described in Chapter 1, when EPA implements statutes requiring risk assessment, the health effects information used in such assessments contributes to improving the understanding of the adverse effects of environmental toxicants, but it does not produce personal benefits to those who participate in the experiments. Air chamber studies of the kind EPA has conducted in the Air Office can occasionally be exceptions to this general rule, because participants who experience angina pain, for example, may benefit from learning more about the circumstances in which they experience such effects. Pesticide-related studies, however, are designed to detect either adverse effects or effects on normal physiologic reactions, or they are designed to study the pharmacokinetics of a chemical in the human body. Secondary benefits might accrue to participants, for example, who receive a comprehensive medical screening evaluation as a condition of participation. However, the possibility of gaining such benefits does not result from the administration of the chemical, and it is not integral to the goals of the study.

Payment for Participation

Paying research participants, which is a common and longstanding practice in the United States, provides a form of personal benefit. Although payments are made in part to compensate participants for the inconvenience they may experience, they also appear to aid in study recruitment. The value assigned to financial compensation of research participants in the risk-benefit analysis has been controversial among ethicists and other experts. The committee did not undertake an in-depth analysis of the issues involved (although Chapter 5 includes a discussion of the role of payment in an individual's decision to participate in a research study). Nonetheless, acknowledging the controversy over how compensation affects the overall risk-benefit assessment seems necessary in light of the near universal practice of paying volunteers for their participation in the third-party studies submitted to EPA that were reviewed by the committee.

With regard to compensation, NBAC's report on *Ethical and Policy Issues in Research Involving Human Participants* illustrates one significant viewpoint. NBAC expresses the concern that treating compensation as a benefit for purposes of the Common Rule's balancing of risks and benefits "would inappropriately skew judgments concerning risks and potential benefits, because nearly any level of research risk could be offset by such gains if they were significant enough—for example, if participants were promised large sums of money for participating in the research" (NBAC, 2001, 74). In light of this concern, NBAC urged that compensation not be considered a benefit for purposes of an Institutional Review Board's (IRB's) weighing of the risks and benefits of a research proposal. This result strikes some as counterintuitive, ignoring the undeniable fact that from the perspective of a prospective participant, compensation can and often does count as a benefit, and even one that may tip the balance in the individual's decision-making process regarding whether or not to participate.[1]

Qualms about the correct treatment of compensation partially reflect an interest in preserving a central feature of both the Common Rule and other statements of principles regarding human research: the requirement that not one but two affirmative judgments must be made in order for research to be designated as appropriate—one by an individual when providing informed consent, and the other by an independent body evaluating risks and benefits, as well as other features of the research protocol. Under this structure, it should be possible for a potential research participant to give informed consent but for the IRB to consider the protocol unacceptable because of its risks. If the amount of compensation could count as a benefit in the IRB's assessment, just as it might play a part in the individual's decision to participate, the two judgments would become difficult to distinguish. The independent assessment contemplated by the Common Rule seems designed to reflect broader social norms regarding acceptable research, norms that cannot be offset by the promise of greater payment to participants.

In the end, the committee did not attempt to resolve definitively the extent to which compensation should be considered a personal benefit for purposes of the independent appraisal of whether a study's benefits justify the risks involved. Committee members did agree, however, that if compensation were the only benefit of an intentional human dosing study, this would be inadequate to justify any risk. Because generally in human studies conducted for EPA regulatory purposes there are no other personal benefits to participants beyond compensation (however that is judged as a benefit),[2] the justification for such studies depends on the presence of sufficient societal benefits to justify the risks.

---

[1]NBAC also expressed concern that high levels of compensation would undermine informed consent by "induc[ing] participants to enroll without carefully considering the risks involved in participation" (NBAC, 2001, 74). See Chapter 5 with regard to this aspect of the compensation controversy.

[2]There may be extraordinary cases in which personal benefits are present, but such studies would have to exhibit some distinctive feature not present in the studies that the committee reviewed.

Societal Benefits

Identifying and assessing the societal benefits of intentional human dosing studies and then comparing those benefits to the risks to participants are a controversial and complex process. In the context of the pesticide program, the Joint Subcommittee of the EPA Science Advisory Board (SAB) and the Federal Insecticide, Fungicide, and Rodenticide Act (FIFRA) Scientific Advisory Panel (SAP) (discussed in Chapter 1) concluded that, in order for such studies to be ethically justified, "the information expected to be gained must promise reasonable health benefits to the individual or society at large," and that even then such studies should be considered only if they meet conditions that the report described as ranging from "rigorous to severe" (EPA, 2000, 3). The requirement that the study should hold the promise of providing health benefits means that the SAB/SAP subcommittee "would not support human dosing that intended to bring about increased allowable residue levels" for a pesticide (EPA, 2000, 26), because no health benefits are achieved when tolerance levels for a pesticide are raised. Two members of the subcommittee of 13 filed a minority report expressing still greater reservations about intentional dosing studies. They contended "that no limited human study will provide information about safe levels of intake of pesticides by humans, especially humans" (EPA, 2000, C-1). Those who signed the minority report apparently concluded that the type of studies the subcommittee had been asked to review could never be conducted ethically.

For reasons explained below, this committee does not agree with the SAB/SAP subcommittee in two important respects. First, the committee believes that environmental as well as health benefits should be considered. Second, studies meeting the six conditions imposed by the regulatory framework on human research noted at the beginning of this chapter, but whose results do not promise health or environmental benefits may be acceptable if (1) there is a sound scientific basis for concluding that the exposure during the study to the chemical being tested will not harm research participants and (2) the study would make an important contribution to the scientific quality of a regulatory decision, whether that decision is to decrease or increase an allowable residue level (which, of course, cannot be known with certainty until the study is conducted).

This conclusion hinges on the committee's determination regarding what constitutes a societal benefit for purposes of evaluating the ethical validity of human dosing studies conducted for EPA regulatory purposes. For many of the same reasons discussed earlier in the context of personal benefits, compensation to participants plays no role in assessing societal benefits. However, the committee identifies two distinct types of societal benefits that might accrue: (1) improving the scientific basis for implementing congressionally mandated regulatory frameworks with all of the community benefits that this implies and (2) human health or environmental benefits that might result from the use of human data in setting regulatory standards.

As noted in Chapter 1, the environmental toxicants over which EPA has jurisdiction under its statutes pose regulatory policy challenges because they produce risks and they are produced for or released as a result of activities that society values. The policy challenge society faces is developing an acceptable means for resolving the clash of interests or values produced by this dilemma. It is, furthermore, a difficult policy challenge, because people disagree over the value of the activity that generates the toxicant, over how essential it is to carry out the activity in a way that generates toxicants, over how much risk is produced, over how to value that risk, and over how all of these considerations should be weighed in the ultimate resolution. Nonetheless, the competing values must be resolved—even as these subjects continue to be debated—and the resolution is created through legislation and legislatively mandated administrative decision processes.

In a functioning democracy, the particular resolution embodied in statutes, regulations, and administrative procedures should be accorded legitimacy, even as efforts may be made by some to change the law. Bringing policy as implemented into closer alignment with policy as enacted, therefore, confers greater legitimacy to government decisions, which is a societal benefit, regardless of whether the result of human testing is to make the regulatory standard more or less stringent. If a different legislative resolution occurs, bringing policy implementation into closer alignment with that different resolution will be what produces a societal benefit. Acknowledging the societal benefit of an improved scientific basis for making decisions obviously does not resolve legislative or public controversy over how the risks of toxicants should be regulated.

A second type of societal benefit consists of benefits to human health or the environment that result from the implementation of a regulatory standard. For example, an air pollution standard that is made more stringent on the basis of human studies provides a health benefit to those who will be protected from the adverse respiratory effects of a pollutant. Because some pesticides contribute to disease control, a risk evaluation that allows the use of such pesticides may produce a public health benefit. Other pesticides do not generate such public health benefits. Recognizing the possible differences in such downstream consequences is a necessary step when comparing the risks and benefits that may result from intentional human dosing studies.

It is important to note that it is not clear whether or how such issues should be considered within the current regulatory framework for human research. Specifically, the Common Rule at 40 CFR 26.111(2) states that "the IRB should not consider possible long-range effects of applying knowledge gained in the research . . . as among those research risks that fall within the purview of its responsibility." It is not apparent from the text of the Common Rule what this language is intended to mean in the context of studies conducted for EPA regulatory purposes. The committee believes that, in the EPA context, considering the benefits associated with the kinds of uses to which tested substances will be put is no less relevant to an IRB review than are the anticipated health-related uses to which a tested pharmaceutical will be put when an IRB is reviewing a drug trial.

The following sections discuss both types of societal benefits—improving the scientific basis for implementing legislation and human health or environmental benefits.

RELIABILITY IN IMPLEMENTING THE CURRENT REGULATORY FRAMEWORK

As a society, we currently employ a variety of approaches to accommodate public health concerns raised by the use of environmental toxicants that are associated with useful activities. Some, such as the Emergency Planning and Community Right to Know Act (42 U.S.C. 116), require sources of pollutants to report the amount of particular harmful substances released into the environment. Some, such as the new source performance standards of the Clean Air Act (42 U.S.C. 85), require sources of pollutants to reduce the release of specific harmful substances to levels attainable through the application of pollution abatement technology that EPA has judged practicable, or best economically achievable, or best available, or that meets some other technology standard established by the statute. Others, such as the ambient air quality standard-setting provisions of the Clean Air Act or the tolerance setting process under the Food Quality Protection Act of 1996 and the Federal Food, Drug, and Cosmetic Act (FFDCA), mandate that levels of particular harmful substances should not exceed the levels judged to be low enough to protect humans from specified adverse health effects. Still others, such as the registration process under FIFRA, require EPA to balance the adverse effects and the beneficial effects of permitting the environmental release of harmful substances.

The advantages and disadvantages of each of these approaches have been debated at great length.[3] Whatever approach Congress has chosen,

[3]For a useful summary, see Office of Technology Assessment, *Environmental Policy Tools* (1995), available at www.wws.princeton.edu/~ota/ns20/alpha_f.html.

a constant has been the need to develop factual information for its implementation, much of which is scientific in nature. Major differences among the diverse approaches include the particular type of scientific information needed and the conclusions that must be reached in order to implement them. Health-based approaches—such as the ambient air quality setting process of the Clean Air Act or the process for setting tolerances for pesticide use on food on the basis of a "reasonable certainty of no harm," and the risk-benefit balancing approaches, such as the licensing process for nonfood use pesticides—require information that relates exposure to the substance to types and levels of harm. In other words, they require some assessment of the risks associated with the substance. Where the risks to humans are among those that need to be assessed—as they are in the cases of the Clean Air Act and FIFRA—then information that could predict potential human responses to exposure is relevant to that risk assessment, especially in the hazard identification and dose-response assessment components. (The general risk-assessment framework is described in Chapter 1.)

Those who assess risk try to provide information to risk managers that rests on reliable science, information that typically is drawn primarily from animal toxicity studies. However, except in cases in which the specific risk of concern is directly measurable in humans, nothing guarantees that science at any point produces correct answers. As the U.S. Supreme Court has recently remarked, "it would be unreasonable to conclude that [scientific conclusions] must be 'known' to a certainty; arguably, there are no certainties in science" (*Daubert* v. *Merrell Dow Pharmaceuticals, Inc.*, 509 U.S. 579, 590 [1993]). In many cases, again in which the effect of concern cannot be directly measured in human studies, all that science can provide is a determination supported by a broad segment of the scientific community—where one exists—that a particular approach or finding represents the best understanding at any particular point in time.

In many instances, reputable science will be unable to generate all the findings necessary to give risk managers a completely science-based set of findings on which to predicate the public health and welfare decisions that must be made. For almost all risk assessments involving toxicants, current scientific knowledge is insufficient to reach a definitive conclusion with regard to some of the questions such assessments raise, because in almost all cases, human toxicity, especially long-term toxicity, is difficult or impossible to study directly.

For example, scientific studies could establish that a toxicant causes malignant tumors in several animal species. What do these results say about the carcinogenic potential of the substance in humans? The animals were exposed to high levels of the substance for a prolonged period. What do these high-exposure results say about the ability of the substance to produce malignant tumors at the relatively low and often intermittent doses to which humans are exposed? Is there a level of exposure below which the substance does not have any potential to cause adverse effects? Or can we describe with certainty an exposure that produces a specific level of risk that we would consider acceptable? These kinds of questions cannot be answered with certainty by today's science. Yet some judgment about the answers must be reached by agency decision makers who have been charged with making regulatory determinations based on an assessment of risks.

Making a regulatory judgment (including a decision to do nothing), cannot be delayed until science makes all the predicate findings. Some determination of how to weigh useful activities and the risks they create must occur now, even though it may need to be changed later. Therefore, public policy decisions regarding these questions will have to be made even in the presence of important gaps in knowledge that cannot be filled by science at this time. These decisions will necessarily involve both scientific findings and judgments about how the gaps in knowledge should be filled.

The 1983 National Research Council report, *Risk Assessment in the Federal Government: Managing the Process*, identified 51 different places in a routine risk assessment where the exercise of judgment can be required to bridge a gap between what science is currently prepared to accept as a valid finding and the next step in the analytic process that constitutes the risk assessment (NRC, 1983, 33-37). The report referred to these gap-filling judgments as "inferential bridges." EPA refers to them as "default assumptions"—that is, the risks that will be presumed to exist in the absence of other data. These assumptions reflect the agency's current thinking regarding how a question should be answered in the absence of additional scientific evidence indicating that a different answer would be better. Thus, for example, in the absence of scientific evidence showing that there is a safe, nonzero level of exposure to a human carcinogen, EPA's default assumption is that there is no such level. Accordingly, any human exposure to such a substance is assumed for regulatory decision-making purposes to create some risk of contracting cancer (NRC, 1994).

These default assumptions allow risk-management decisions that are based on human risk assessments to be made under conditions of uncertainty, an unavoidable necessity. At the same time, the presence of such default assumptions produces another imperative. When possible, the default assumptions should be adjusted and eventually replaced with findings or judgments that are rooted in improved scientific understanding. This imperative is implicit in the commitment to use the best available science in making risk-management decisions. The committee determined that this commitment is sound, subject, as discussed below, to the equally important and often overriding need to protect participants in research. Such a commitment to the use of the best available science has been made and consistently reaffirmed by all three branches of government—executive, legislative, and judicial—and accordingly it should be considered by EPA in assessing the benefits of all scientific results, including those involving intentional human dosing studies.

Executive Order 12866, issued initially by President Clinton in 1993 and revised by President Bush in 2003 (without changes to the relevant portions), directs each administrative agency to "base its decisions on the best reasonably obtainable scientific, technical, economic, and other information concerning the need for, and consequences of, the intended regulation" (E.O. 12866 §(b)(7)). EPA has a similar longstanding and publicly stated commitment to using the best available scientific information. In 1991, the agency issued a mission statement that included the commitment to ensure that "national efforts to reduce environmental risk are based on the best available scientific information communicated clearly to the public" (EPA, 1991). A year later, an expert panel on science at EPA reiterated this commitment in a report to Administrator William Reilly (EPA, 1992). In 1994, a policy guideline from EPA Administrator Carol Browner stated that "EPA strives to ensure that the scientific and technical underpinnings of its decisions meet two important criteria: they should be based upon the best current knowledge from science, engineering and other domains of technical expertise; and they should be judged credible by those who deal with the Agency" (EPA, 1994). The agency's current mission statement also commits the agency to ensuring that "[n]ational efforts to reduce environmental risk are based on the best available scientific information."[4]

Default assumptions in risk assessments are needed in areas in which science has not progressed sufficiently to provide an answer to a question that is a necessary part of a risk assessment. When an answer to such a question becomes available, however, the general imperative of using the best available science implies that this answer should replace the default assumption. Numerous specific pronouncements by the agency regarding default assumptions bear out the desirability of replacing default assumptions with scientific results. For example, the Draft Water Quality Criteria Methodology Revisions state that "When adequate data are available they are used to make accurate exposure predictions for the population(s) of concern. When this is not possible, a series of qualitative alternatives is proposed using less adequate data or default assumptions

[4]Available at www.epa.gov/history/org/origins/mission.htm.

that allow for the inadequacies of the data while protecting human health" (EPA, 1998a).

The Guidelines for Neurotoxicity Risk Assessment are similar, stating that "default assumptions should not be applied indiscriminately. First, all available mechanistic and pharmacokinetic data should be considered. If these data indicate that an alternative assumption is appropriate or if they obviate the need for applying an assumption, such information should be used in risk assessment." (EPA, 1998b). Finally, EPA's Proposed Guidelines for Carcinogen Risk Assessment state that "EPA's 1986 guidelines for cancer risk assessment . . . were developed in response to [the Red Book]. The guidelines contained a number of default assumptions. They also encouraged research and analysis that would lead to new risk assessment methods and data and anticipated that these would replace defaults" (EPA, 1996).

There is strong evidence that Congress has consistently shared with the executive branch the view that when science is to be relied on to supply information pertinent to a regulatory decision, the best available science should be employed. Sometimes it has stated this view explicitly, as in the 1996 amendments to the Safe Drinking Water Act. There, Congress has provided that "to the degree that an [EPA] action is based on science, the Administrator shall use:

> (i) the best available, peer-reviewed science and supporting studies conducted in accordance with sound and objective scientific practices; and
>
> (ii) data collected by accepted methods or best available methods (if the reliability of the method and the nature of the decision justifies use of the data) (42 U.S.C. §300g-1(b)(3)(A)).

Examples of statutory language requiring the use of the best available science can be found in older statutes as well. The Asbestos School Hazard Abatement Reauthorization Act of 1989 requires that when EPA provides information to schools about the hazards of asbestos, "[s]uch information or advisory shall be based on the best scientific evidence. . ." (15 U.S.C. §2643).

Congress's commitment to using the best science available is not limited to actions taken by EPA. For example, the Endangered Species Act was amended in 1978 to include the instruction to all federal agencies that "each agency shall use the best scientific and commercial data available" in ensuring that any action by an agency will not threaten the existence of an endangered species (16 U.S.C. §1536(a)(2)). Dating back to 1970, the Occupational Health and Safety Act contains provisions regarding the protection of workers from exposure to toxicants that require protective measures to be based on the "best available evidence" (29 U.S.C. §655(b)(5)).

Besides these statutory instructions, the scientific advisory panels and peer review procedures established by law under many of the statutes EPA administers provide further evidence of Congress's appreciation of the value of improving the quality of scientific findings that inform agency decision making. Under FIFRA, for example, the Administrator is to "solicit from the [scientific] advisory panel comments, evaluations and recommendations for operating guidelines to improve the effectiveness of scientific analyses made by personnel of the Environmental Protection Agency that lead to decisions by the Administrator in carrying out the provisions of this subchapter" (7 U.S.C. §136w(d)(1)).

Not all of the statutes EPA administers explicitly invoke the use of the "best available science," or something equivalent in statutory language, but these illustrations show that many statutes reflect the conviction that when scientific judgments are called for, better science is preferred.[5] There is no reason to believe that when statutory language does not contain such explicit language, the presumption that the best available science should be employed should be any different. Regardless of the statute and the science involved, improvements in the accuracy and reliability of the science improve the quality of information that is relied on for making ultimate regulatory decisions. It is hard to imagine that Congress would not consider the improvement of the quality of scientific information to be a benefit to the regulatory processes that it has asked EPA to implement. The committee thus concludes that when Congress has enacted regulatory processes that rely on science, improving the science those processes employ serves to implement the resolution of competing interests.

[5]Determining what constitutes best available science is a decision for the agency. A reviewing court will accord the agency considerable deference regarding these decisions. Court challenges to scientific findings made by the agency have involved situations in which the available science is inconclusive or conflicting or of disputed validity. In these cases, EPA must provide an explanation of how it reached its conclusions and why it declined to follow studies upon which it did not rely, but the courts generally defer to these explanations. The committee was aware of only one reported decision in which an EPA action was reversed because the agency relied upon inferior science—and in that case the agency itself conceded that the best available science supported a different result. In that case, the court enforced the Safe Drinking Water Act (SDWA) requirement that the agency employ the best available science. See *Chlorine Chemistry Council* v. *EPA*, 206 F.3d 1286 (DC Cir 2000). The Chlorine Chemistry Council case reversed a decision to retain a maximum contaminant level goal (MCLG) of zero for chloroform despite EPA's awareness of reliable scientific studies suggesting that a nonzero level was justified. The case arose in the unusual circumstance in which EPA had conceded the validity of findings pointing to a nonzero MCLG, but had declined to rely upon those findings because the scientific studies had not been completely reviewed through the statutory scientific advisory and peer review processes. The agency had stated an intention to rely upon the newer findings as soon as these review processes had been completed, but the court held that the SDWA did not permit the agency to wait.

Balancing the interests of the various parties affected by EPA's statutory requirements can lead to the development of regulations that effectively establish legal rights and responsibilities. Parties can correctly insist that the rights and responsibilities that are ultimately established and enforced by EPA through its regulations implement the public policy that Congress has enacted. In our democratic system of government, Congress codifies a particular balance when it writes binding law, and that balance should be observed by administrative agencies until it is changed.

Even in cases where the agency employs elements of discretion in fine-tuning the ultimate regulations, that discretion should be based on the best relevant and available understanding of the information Congress has directed the agency to take into account. In the case of risk assessments and the regulatory decisions that employ them, this information includes the scientific components of the assessment. The more accurate the science-based components of the regulatory systems EPA administers under these statutes, the better informed EPA's exercise of discretion will be.

In addition to its stated general preference for replacing default assumptions with scientific findings, EPA has expressed a specific preference for supplementing animal data with human data when conducting human risk assessments. EPA has said that it looks to human data whenever possible in completing human risk assessments: "If adequate human studies (confirmed for validity and applicability) exist, these studies are given first priority in the dose-response assessment, and animal toxicity studies are used as supportive evidence" (EPA, 1989). Often, such data can be obtained from epidemiological studies, which do not involve the intentional dosing of research participants, but rather evaluate the effects of exposures that have occurred in an occupational setting or because of the peculiarities of a specific geographical setting.[6] Regardless of the origins of such human data, "risk assessments based on human data have the advantage of avoiding the problems inherent in interspecies extrapolation" (EPA, 1993).

The default assumptions that are of particular relevance to the issues raised by third-party intentional human dosing studies are those that bridge gaps between animal results and estimates of effects in humans. In the context of FIFRA, for example, EPA has routinely divided the calculated "safe" dose for animals by a factor of 10, to account for the possibility that humans are more sensitive to the substance being tested than are

[6]This quotation from the Risk Assessment Guidance refers specifically to epidemiological data.

the animal species. Third-party submitters of human dosing studies have been particularly interested in modifying this default assumption by introducing data obtained directly from human studies.

The benefits to the regulatory process of improved science are generally accepted without question in areas of risk assessment that do not involve the deliberate exposure of humans to toxicants. For example, the fate and transport studies central to defining the nature and extent of human exposure require an understanding of how substances released into the environment move in that environment, interact with other substances, and eventually come into contact with humans, whether through dermal contact, inhalation, or ingestion. Answering these questions involved in fate and transport studies involves applying knowledge in fields such as hydrology and chemistry. There is little controversy regarding the idea that improving the accuracy and reliability of the science benefits the risk-management process by providing the best answers to scientific questions that can be provided at the time.

The critical difference between improving the exposure assessment component of a risk assessment through better fate and transport models and improving the dose-response component of that assessment through human studies is not that the first supplies a benefit to the regulatory process and the second does not. Both provide benefits in the form of better estimates to use in the risk-management process; however, this certainly does not mean that it should be federal policy to pursue either of these benefits indiscriminately. To say that a piece of information supplies a benefit is not the same as saying that we should acquire the information regardless of the costs. A major commitment involved in ensuring the ethical treatment of research participants is being prepared to reject research that would produce beneficial information if that research exposes humans to unjustified risks. The difference between improving an exposure assessment and improving a dose-response assessment is that the former typically does not expose humans to health risks, while the latter, if it is to be accomplished by experimentation directly on humans, potentially does. This difference obviously has tremendous significance and a profound effect on how one should approach evaluations of those studies. In terms of the risk-benefit calculus that would be applied to judging the ethical acceptability of a human study, the way to take this difference into account is first by making a careful determination of what the risks are and then weighing those risks against any benefits that might result from the study.

The committee concludes that it is a matter of established and sound public policy that the use of the best available science—including the replacement of default assumptions with reliable scientific information—constitutes a societal benefit.

As discussed earlier, the Common Rule requires that there should be not only an expectation of benefit resulting from the proposed research, measured for present purposes by the "importance" of the knowledge to be gained, but that risks to participants should be considered as well and that these risks should be "reasonable" in relation to the importance of the knowledge. This standard requires that the risks and benefits of a study be evaluated and then compared.

Building on these principles, the following section provides the committee's perspective on how the balancing of risks and benefits should be approached with respect to human studies conducted for EPA regulatory purposes.

BALANCING RISKS AND BENEFITS

Of the three basic ethical principles governing the protection of research participants—respect for persons, beneficence, and justice—beneficence is the one that, in the context of this report, requires the greatest exercise of subjective judgment with the least amount of guidance from established policy or precedent. Informed consent (respect for persons) and the fair distribution of the benefits and burdens of human research (justice) both are important and challenging, but it is possible to delineate reasonably objective decision rules to guide their application.

Beneficence is the ethical principle that requires considering the well-being of the research participant and ensuring that possible risks are minimized and that any risks that remain are justified by the potential benefits of the research (National Commission, 1979). Beneficence thus requires a subjective balancing judgment. Moreover, in the context of human research conducted to inform EPA's regulatory decision making, beneficence requires balancing anticipated risks to the participant against potential benefits to society in order to assure that the risks are justified by the benefits. To paraphrase the Common Rule, the risks must be "reasonable" in relation to the importance to society of the knowledge produced by the research. There are no formulas for determining whether a risk to an individual is justified by a benefit to society.

Independent review of human research, such as is conducted through local IRBs, is essential to ensuring that all three of the key ethical principles are being followed. In the case of clinical research on therapeutic products, IRBs have considerable experience in balancing risks and benefits and are also familiar with certain kinds of studies in which the benefit does not accrue directly to study participants, such as pharmacokinetic (PK) and pharmacokinetic-pharmacodynamic (PK/PD) studies, other mechanistic studies, and Phase 1 studies. In these cases, the kind of information to be obtained and its usefulness are relatively familiar. In the cases addressed in this report, however, concerning human studies conducted to inform EPA's regulatory decision making, most IRBs have little or no experience in weighing the kinds of benefits that might arise against the risks. This is one reason why the committee recommends later in this report (see Chapter 6) that there should be a role for a centralized review body operating under EPA's auspices to review human studies conducted for EPA regulatory purposes.

In the next section, the committee provides an overview of the kinds of risks and benefits that such studies may present. It also provides some perspectives on how the risks and benefits might be balanced to determine whether a study comports with the principle of beneficence.

Assessing the Risks

The Common Rule requires investigators and IRBs to identify, analyze, and assess risks, and investigators to disclose risks to potential research participants. The term "risk" refers to the probability of a harm occurring and includes consideration of both the magnitude of a particular harm and the probability of its occurrence. Because both the risks and the benefits of research are not known in advance and can only be projected or predicted, the proper comparison is not between risks and benefits but rather between anticipated risks and potential benefits.

Under the Common Rule, IRBs evaluating research protocols are required to (1) classify risks (as minimal or greater than minimal), (2) ensure that "risks to subjects are minimized," and (3) determine that risks are reasonable in relation to probable benefits to research participants and/or the "importance" of the reasonably expected knowledge. Each task poses important challenges for IRBs. The first two are discussed here, and the third—the balancing of risks and benefits—is discussed later in this section.

The distinction in the Common Rule between minimal and greater-than-minimal risk provides a sorting mechanism that enables IRBs to attend more closely to protocols that involve greater risks. A classification as minimal risk is a necessary (but not sufficient) condition for a protocol's expedited review, rather than full convened IRB review, and for a waiver or modification of the elements of informed consent or of the documentation of informed consent. Another category, "a minor increase over minimal risk," has been adopted by the Department of Health and Human Services (DHHS) and by the U.S. Department of Education for research involving children (Subpart D of the DHHS version of the Common Rule, 45 CFR 46).

The minimal risk standard encompasses studies whose risks are so low that customary IRB review and even some elements of informed consent can be bypassed. Most studies that qualify as minimal risk under the Common Rule involve no active intervention affecting the research participant—that is, they are observational or epidemiological rather than invasive. A study of postexposure pesticide levels might belong in the minimal risk category. Although the language of minimal risk is widely used in the United States and in international discussions, its interpretation varies, especially in cases that involve some active intervention. According to the Common Rule, "Minimal risk means that the probability and magnitude of harm or discomfort anticipated in the research are not greater in and of themselves than those ordinarily encountered in daily life or during the performance of routine physical or psychological examinations or tests" (40 CFR 26.102(h)(1)). Even with this definition as a guide, minimal risk is not construed in consistent ways among federal agencies or by different IRBs. In view of these inconsistencies, NBAC proposed that:

> IRBs should use a standard related to the risks of daily life that are familiar to the general population for determining whether the level of risk is minimal or more than minimal. *The standard should not refer to the particular risks encountered by particular persons or groups* [emphasis added]. It should refer, therefore, to common risks—for example, driving to work, crossing the street, getting a blood test, or answering questions over the telephone (2001, 83).

NBAC made this distinction because people who face inherently risky situations, by virtue of, for example, illness or occupation, should not be allowed to face higher risks in research than others, except in exceptional circumstances (e.g., compassionate use of experimental treatments in the terminally ill).

The committee finds the experience with the minimal risk concept in the context of clinical research uninformative for purposes of assessing the ethical validity of the types of human studies most likely to be conducted for EPA regulatory purposes. Even though some of the intentional dosing studies conducted for these purposes pose no identifiable risk to participants, the committee is reluctant to consider any toxicant dosing study a minimal risk study within the meaning of the Common Rule. Importantly, the committee concludes that any human dosing study conducted for EPA regulatory purposes, regardless of how safe it may appear to be, and even if it could be judged by some to pose minimal risk under the Common Rule, should be reviewed both by an IRB and by the Human Studies Review Board recommended in Chapter 6. This will ensure that the health of participants is in fact protected. It also reflects the need for careful review to ensure that a proposed study provides the specialized form of societal benefit—improving the scientific quality of regulatory decision making—potentially associated with studies conducted for

EPA purposes. Thus, even if a human dosing study conducted for EPA regulatory purposes could be deemed to pose minimal risk, that finding would not, under the committee's recommendations, have the practical consequences it has for the more typical human research study evaluated under the Common Rule.

For these reasons, in describing the range of risks posed by the human dosing studies addressed in this report and how those anticipated risks might be balanced against potential benefits, the committee does not use the terminology of "minimal risk" or "minor increase over minimal risk." Rather, the committee uses other terms that describe the anticipated risk or lack thereof, as discussed below, but will do so free of the implications the term "minimal risk" carries as applied in other settings under the Common Rule. This approach also is intended to better reflect the nature of the human dosing studies that are generally conducted for EPA purposes and the range of possible human responses to chemical exposures.

Exposure to any chemical substance, whether of natural or industrial origin, can cause alterations of many types in the biological structures and functions of living organisms, including humans. These alterations vary among chemicals and also with the conditions of exposure (with conditions referring to the magnitude, duration, and route of exposure). In addition, for most chemicals there are ranges of doses outside of which no biological change in structure or function can be detected using the best available scientific technology. For example, as discussed in Chapter 3, PK studies of toxicants, which are intended to document how a chemical is normally metabolized by the body rather than to elicit any response or alteration, are often conducted at doses that are not expected to cause any significant or even detectable alteration in biological structures or functions. In other intentional human dosing studies involving pesticides, the objective is to elicit some biological response and to identify a dose at which the response did not occur. In such studies, the maximum studied dose at which no biological changes can be observed (always relative to a control group) is referred to as the "no observed effect level," or $NOEL_{HU}$, but to determine the $NOEL_{HU}$ rigorously it is necessary to find the dose at which the effect is seen, a "lowest observed effect level," or $LOEL_{HU}$. In other studies the effect investigated could be relatively mild but nonetheless undesirable (i.e., perceived as an adverse event). The committee saw no examples of studies conducted for EPA that were intended to provoke more serious adverse effects. However, one study did produce an effect larger than expected at the midrange of the intended dosing schedule and was stopped.

With this as background, and based on its review of the kinds of human dosing studies that have been and are likely in the future to be submitted to EPA, the committee identifies three categories of anticipated risk associated with such studies.

The first category of risk includes studies that pose *no identifiable risk to participants.* This category includes PK studies conducted at low doses that delineate uptake and disposition of a chemical and its metabolites, but are expected, based on extensive previous testing in animals, to have no biological effect on the participant, as discussed in Chapter 3. Although it is not possible to prove the total absence of risk with absolute certainty, low-dose PK studies of the kind noted here and discussed in Chapter 3 are typically conducted at levels far below those that have been or would be judged safe under the legal safety standard in FFDCA for pesticide residues in food and are as close to being risk free as any human dosing study can be.

The second category of risk includes studies that elicit a biological response, but ones that are not in any way adverse to the participant, such that, on the basis of ample scientific evidence, experts would conclude that there is *a reasonable certainty of no harm to study participants.* This category includes the PD studies discussed in Chapter 3 in which the observable changes serve as indicators or biomarkers of exposure, but are immediately reversible upon cessation of exposure and would be expected to have no consequence to the health of the individual experiencing them. Changes, for example, in cholinesterase activity in blood would be rapidly reversible and at low exposure would not be associated with any adverse effect. Detectable but clinically insignificant changes in blood pressure or heart rate in normotensive individuals would similarly be considered nonadverse and are often categorized as indicators of exposure to a chemical, rather than as evidence of toxicity. In some cases, it is clear that those biological changes, while not adverse in themselves, are sensitive indicators of a process that would be adverse if the effect were greater (e.g., from a larger dose) or if dosing were prolonged. Cholinesterase inhibition studies on *organophosphate* (OP) pesticides are examples of such cases, as discussed further in Chapter 7.

The third category of risk includes studies at doses that elicit a biological response in either the structure or function of the organism that is potentially detrimental to health, or adverse, and thus *poses an identifiable risk to study participants.* This category theoretically encompasses a wide range of risks. It includes potential risks posed in the PD studies discussed in Chapter 3 that have adverse effects that are non-trivial but transitory and expected ultimately not to be harmful to the study participants, in the sense of causing any lasting impairment or pain. Examples of such transitory, non-trivial symptoms that are adverse but not ultimately harmful include headache, nausea, or temporary irritation to the eyes or airways. These are symptoms sometimes seen in air pollution studies.

Studies whose adverse effects are not transitory and thus may result in lasting harm are also included in this category, but the committee is unaware of human dosing studies conducted for EPA regulatory purposes that were anticipated to result in lasting harm to participants. Such studies are clearly not allowed. Even when the adverse effect is transitory, however, such studies pose an identifiable risk of immediate harm and, especially if conducted in vulnerable populations, pose a significantly greater risk of unexpected lasting harm than the studies in the first two risk categories.

Assessing the Benefits

Any human dosing study, regardless of its risk category, must have a useful purpose and convey some benefit to the participants and/or society. As discussed earlier, the committee concludes that under the risk-benefit balancing required by the principle of beneficence and the Common Rule, personal benefits to participants are insufficient by themselves to justify human dosing studies conducted for EPA regulatory purposes. This means that risks to participants imposed by human dosing studies must be justified by the societal benefits that are anticipated to come from a successful study, if they are to be justified at all. In this respect, human dosing studies are similar to Phase 1 drug trials.

The committee also concludes that in order to generate societal benefits at all, human dosing studies must (1) be performed in a context in which there is a clearly defined regulatory objective and a critical, unanswered question or other compelling scientific need that cannot be satisfied with animal data and (2) be designed with the requisite statistical power and other design features required to meet that regulatory objective and scientific need. These are threshold requirements that any human dosing study must meet. The steps that study sponsors must take to satisfy this threshold test are discussed in Chapter 3.

Studies that satisfy this threshold test have the ability to improve the accuracy of EPA's regulatory decision making. For the reasons discussed earlier in this chapter, the committee concludes that improving the accuracy of the science employed in regulatory decisions constitutes a societal benefit. Beyond this minimal benefit, however, human dosing studies also can generate different kinds of societal benefits as well, depending on the nature of the scientific question a study seeks to answer, the uses to which study results may be put, and the consequences that may flow from those uses. Thus, just as there is a spectrum of risk categories into which human dosing studies might fall, there is a spectrum of potential societal benefits, which can be categorized roughly as follows.

The first benefit category is the one outlined above in which the study *provides improved accuracy of EPA decision making* but conveys no other societal benefit in terms of better protecting human health or the environment. This benefit category provides the minimum benefit required to justify a human dosing study. The third-party studies that have been recently submitted to EPA's pesticide program aimed at identifying the $NOEL_{HU}$s and $LOEL_{HU}$s for specific OP pesticides are the most prominent examples of studies that may provide benefits in this category. In these cases, extensive animal testing has established that the critical determinant of toxicity (and risk) for the pesticide is cholinesterase inhibition. Thus, use of data on cholinesterase inhibition from humans, assuming they derive from properly designed and executed studies, would improve the scientific accuracy of EPA's risk assessment. Although such studies conceivably could demonstrate that humans are more sensitive than animals and that the Reference Dose (RfD) derived from human data is lower than one based on animal data, the interest of the study sponsor is to increase the RfD and thus allow for greater use of the pesticide based on a more scientifically accurate risk assessment.

Determining the implications of the benefits from such studies was challenging to the committee, as they could result in the reduction or elimination of the uncertainty factor, which could produce a less stringent regulatory standard. The committee concludes that when a study improves scientific accuracy relevant to regulatory decision making but generates no health or environmental benefits, the benefit of the improved scientific accuracy of decision making can justify the intentional exposure of humans only to the lowest two categories of risk, as outlined above. This means that such studies must pose no identifiable risk because they elicit no biological response or, in the case of studies that elicit a nonadverse response (such as a nonadverse change in a biomarker), there must be a reasonable certainty of no harm to study participants based on a careful review of an adequate body of scientific evidence. The OP-related human dosing studies submitted to EPA's pesticide program have measured cholinesterase inhibition as a biomarker of exposure and potential toxicity, rather than as a toxic endpoint per se. As explained in Chapter 3, an independent review could conclude that there is a reasonable certainty in such studies of no harm occurring to study participants.

Recommendation 4-1: Value of Studies That Seek to Improve the Accuracy of EPA's Decisions But Do Not Provide a Public Health or Environmental Benefit

EPA should consider a human dosing study intended to reduce the interspecies uncertainty factor (for example, a study of a biomarker such as cholinesterase inhibition) as conferring a societal benefit only if it was designed and conducted in a manner that would im-

prove the scientific accuracy of EPA's extrapolation from animal to human data. Because the anticipated benefit would not be as great as that conferred by studies intended to provide a public health or environmental benefit, the study could be justified ethically only if the participants' exposure to the pesticide could reliably be anticipated to pose no identifiable risk or present a reasonable certainty of no harm to study participants.

The corollary of this recommendation is that a human dosing study on a chemical toxicant that poses an identifiable risk to study participants, even if it involves a transitory adverse effect, can be justified only if the study also provides a benefit to public health or the environment beyond the improvement of the scientific accuracy of the risk assessment underlying EPA decision making about that chemical.

Recommendation 4-2: Value of Studies That Seek to Provide a Potential Public Health or Environmental Benefit

An IRB should be properly constituted to be able to consider whether a study has the potential of providing a clear health or environmental benefit to the community. Such studies could be acceptable even if they involved a somewhat higher level of risk than that posed by studies for which there is no identifiable risk or for which there is a reasonable certainty of no harm. No study is ethically justifiable if it is expected to cause lasting harm to study participants.

There are a number of ways in which a human dosing study could provide benefits to society beyond the minimum benefit of improving the accuracy of regulatory decision making, including the following:

- **The study results in a more stringent regulatory standard.**

Human dosing studies that are reasonably expected to result in more stringent permissible limits for chemicals in the environment or food supply not only improve the scientific quality of the regulatory decision by substituting the more relevant human data for animal data, they also confer a potential public health benefit. Such studies require a risk-benefit balancing, as discussed below, and might be acceptable even if they involved risks somewhat greater than those involved in studies that provide the minimum benefit of improving the scientific accuracy of EPA regulatory decision making.

- **The study enables EPA to adopt a public health measure it otherwise could not adopt.**

Outside of the pesticides setting, EPA itself has from time to time looked to intentional human dosing studies in its air and water pollution programs, where EPA must marshal the evidence required for risk assessment. In the case of air pollutants, for example, intentional human dosing studies under controlled conditions may be the only or best way to reliably estimate the dose-response relationship in humans. Evidence from such studies could be needed to enable EPA to set standards it might not otherwise have been able to set or to set standards that are more fully protective of public health. In such cases, the knowledge derived from the study would have the important societal benefit of improving the population's health. The magnitude of the benefit would depend on the importance of the risk being addressed by the standard and how critical having human data would be to setting the standard at a level that protects health.

- **The study supports approval of a product that protects public health.**

Pesticidal products that are used to control or eliminate disease vectors (such as mosquito or tick control agents) can confer important health benefits to society. If human research were required to understand the risks posed by such products and thus support their regulatory approval by EPA, such research would provide an important health benefit. As in the previous category of benefits, the size of the benefit would depend on the risk being addressed and the importance of the study.

- **The study improves the scientific accuracy of risk assessment for a class of chemicals and/or EPA decisions.**

Studies may have consequences for scientific knowledge that extend beyond the making of any single regulatory decision. Such consequences could occur, for example, if a human study revealed information about the proper extrapolation of animal results to humans that could be applied to an entire class or category of substances, such as the OP pesticides, that operate through a common mechanism of toxicity. By expanding the scope of the benefit to a larger class of EPA decisions without increasing the number of study participants, a study can provide benefits beyond those provided by one whose relevance does not extend beyond a single regulatory decision.

FINDING THE BALANCE

As noted at the beginning of this chapter, determining whether the principle of beneficence has been satisfied requires balancing the anticipated risks to study participants against the anticipated benefits of the study to society. The risks to participants must be reasonable in relation to the societal benefit. In the words of the Common Rule, the risks must be reasonable in relation to the importance of the knowledge that may reasonably be expected to result (40 CFR 26.111 (a) (2)). In the EPA context, if an intentional human dosing study does not have a clearly defined and important regulatory purpose and is not designed adequately to both achieve that purpose and minimize the risks to participants, the study should not be conducted, as such studies needlessly expose humans to health risks. If these threshold requirements are satisfied, risks and benefits can be balanced.

Although the preceding discussion of benefits sheds some light on the judgments that are required to strike an appropriate balance between risks and benefits in the regulatory contexts EPA confronts, the committee recognizes that the balancing requires judgment and that there are no clear rules or formulas. This is why careful independent review of proposed human dosing studies is essential.

At the extremes, the risk-benefit balancing judgment may be relatively easy. In the case of a PK study on a well-tested chemical with established "safe" levels of exposure set through the regulatory process—one that is conducted at dose levels well within the established safe level—there may be no identifiable risk, and the study could be justified if it meets the minimum test for benefits discussed above. Such a study would probably be acceptable if it met a clearly defined regulatory need for the best available scientific evidence. At the other extreme, a study in a medically vulnerable population (e.g., children) that has the potential to cause adverse responses and whose potential to cause lasting harm is uncertain poses risks that would be difficult for potential benefits to outweigh, unless perhaps the substance being tested provided significant health benefits to the study participants or to the class of individuals to which they belong.

The cases between the extremes will be more difficult to evaluate. For example, in an air pollutant air chamber study intended to improve the scientific basis for and health protectiveness of a regulatory standard, how much risk is it reasonable to impose on healthy adults? How great would the potential benefits of the study have to be to justify exposing individuals with impaired pulmonary function? An IRB or other review body would need to consider how important the study would be to the establishment of the standard and whether the risk is reasonable in relation to the societal benefit.

These examples illustrate an important point: Assessing the risk-benefit balance in the case of human studies conducted for EPA regulatory purposes requires careful review and special expertise that is not always available to IRBs. It requires balancing risks to participants against benefits to society that lie in the realm of improved regulatory decision making or in broad public health or environmental impacts. It involves making risk assessments and safety judgments about chemicals that require access to data and expertise that reside at EPA and few other places. It is for these reasons that the committee recommends, in Chapter 6, that local IRB review of human studies conducted for EPA regulatory purposes be supplemented with review by a central, EPA-managed body that has the requisite expertise and that will be publicly accountable for decisions on the ethical acceptability of such studies.

SUMMARY

Weighing the risks and benefits that might arise in human experiments is a critical and particularly difficult element of the ethical evaluation of such studies. Potential or anticipated benefits from studies involving humans can be divided into two broad types—personal and societal. Personal benefits are those that may accrue to an individual by virtue of participation in the experiment. Few intentional human dosing studies promise personal gain. Societal benefits are those that accrue to the society as a whole, or to groups within society, by virtue of the application of the scientific results of the study. This calls for an especially cautious approach in applying general principles and in evaluating, in particular cases, whether the rights and welfare of participants have been adequately protected.

The committee assumes that human studies conducted for EPA regulatory purposes do not confer personal benefits on study participants. This means that the risk-benefit balancing required under the principle of beneficence depends on the evaluation of a societal benefit. Although the volunteer's compensation for participation can be considered a personal benefit at one level, it is properly excluded from the risk-benefit balance for reasons discussed elsewhere in this report.

Benefits do accrue to society, however, when science improves the accuracy of regulatory decisions, including the replacement of default assumptions with reliable scientific information. In the words of the Common Rule, such scientific information has "importance" that should be considered in weighing whether a study is ethically justified. This conclusion is only the starting point, however, for the ethical analysis of human studies. In particular, only risks that are commensurate with this minimal societal benefit can be justified unless additional social benefits also are present. In the case of human dosing studies that provide no further public health or environmental benefits, the committee concludes that such studies that pose no identifiable risk to study participants or that present a reasonable certainty of no harm, based on a careful review of an adequate body of scientific evidence, can be justified under restricted conditions and with appropriate oversight and review regardless of whether the information obtained from the study results in a less stringent or more stringent regulatory outcome.

The committee determined that the analysis and conclusions presented in this chapter could clarify some issues regarding the use of studies that deliberately expose participants to toxicants for EPA regulatory decision making purposes, but it does not pretend to resolve here all of the nettlesome issues that arise from intentional human dosing studies. These ultimately must be resolved through EPA's publicly transparent policy deliberations and through the case-by-case decisions of duly constituted review bodies charged with protecting the interests of participants in particular studies.

REFERENCES

Environmental Protection Agency (EPA). 1989. *Risk Assessment Guidance for Superfund, Vol. 1: Human Evaluation Manual, Interim Final.* EPA/540-1-89/002. Available at www.epa.gov/cgi-bin/claritgw?op-Display&document=clserv:OSWER:1175;&rank=4&template=epa.

EPA. 1991. *Strategic Direction for the U.S. Environmental Protection Agency.*

EPA. 1992. *Safeguarding the Future: Credible Science, Credible Decisions.* EPA/600-9-91-050.

EPA. 1993. *Reference Dose (RfD): Description and Use in Health Risk Assessments, § 1.3.2.2.1.* Background document. Available at www.epa.gov/IRIS/rfd.htm.

EPA. 1994. *Peer Review and Peer Involvement at the U.S. Environmental Protection Agency.* Available at www.epa.gov/osp/spc/perevmem.htm.

EPA. 1996. Proposed Guidelines for Carcinogen Risk Assessment. *Federal Register* 61:17960, 17964.

EPA. 1998a. Draft Water Quality Criteria Methodology Revisions: Human Health; Notice. *Federal Register* 63(157):43755-43828.

EPA. 1998b. Guidelines for Neurotoxicity Risk Assessment. *Federal Register* 63(93):26926-26954.

EPA. 2000. *Comments on the Use of Data from the Testing of Human Subjects: A Report by the Science Advisory Board and the FIFRA Scientific Advisory Panel.* EPA-SAB-EC-00-017. Washington, D.C.: EPA. Available at HtmlResAnchor www.epa.gov/science1/pdf/ec0017.pdf.

National Bioethics Advisory Commission (NBAC). 2001. *Ethical and Policy Issues in Research Involving Human Participants: Vol. 1.* Bethesda, MD: U.S. Government Printing Office.

National Commission for the Protection of Human Subjects of Biomedical and Behavioral Research (National Commission). 1979. *Belmont Report: Ethical Principles and Guidelines for the Protection of Human Subjects of Research.* Washington, D.C.: U.S. Government Printing Office.

National Research Council (NRC). 1983. *Risk Assessment in the Federal Government: Managing the Process.* Washington, D.C.: National Academy Press.

NRC. 1994. *Science and Judgment in Risk Assessment.* Washington, D.C.: National Academy Press.

5

Ethical Considerations in the Review of Intentional Human Dosing Studies

INTRODUCTION

A principal theme of this report is that science and ethics are closely related. As explained in previous chapters, intentional human dosing studies must be both scientifically and ethically justifiable. A human research protocol could be scientifically valid but ethically unacceptable (e.g., because the investigator failed to get informed consent); however, it cannot be ethically acceptable if it does not conform to standards of good research design and conduct.[1] In this sense, sound research design is the first step in developing an ethically acceptable protocol. For these reasons, scientific and ethical considerations need to be integrated in the review and evaluation of research involving humans (IOM, 2003). In addition to meeting standards of scientific validity, as discussed in Chapter 3, intentional human dosing studies also must pass a rigorous risk-benefit analysis, which itself involves both science and ethics, as discussed in Chapter 4.

This chapter addresses the ethical considerations that remain after determining that a research protocol is scientifically valid and that its

[1] A report from the Environmental Protection Agency (EPA) Joint Science Advisory Board and the Federal Insecticide, Fungicide, and Rodenticide Act (FIFRA) Scientific Advisory Panel (SAB/SAP) noted the following: "Bad science is always unethical; research protocols that are fundamentally flawed, such as those with sample sizes inadequate to support reasonable inferences about the matter in question, are unjustifiable" (EPA, 2000, 2).

probable benefits outweigh its risks. These include voluntary informed consent; fair selection and recruitment of potential research participants, including fair payment for their participation; and compensation, including the provision of medical care, for research-related injuries.[2] The chapter concludes with recommendations regarding whether and, if so, how the Environmental Protection Agency (EPA) should use ethically tainted data.

The aim of this chapter is to formulate standards of ethical acceptability of intentional human dosing studies. Because some standards identify a minimum that must be met in any such study, while others point to ideals that should guide such research, it is important to distinguish what is ethically unacceptable from what falls short of ethical ideals.

Federal regulations would not be applicable to many third-party intentional dosing studies, because although EPA has accepted the Common Rule, which governs the research that it conducts or funds, this rule does not apply to privately funded toxicant research. Also, EPA has not adopted Subpart B (fetuses, pregnant women, and human in vitro fertilization), Subpart C (prisoners), or Subpart D (children). In addition, although the Common Rule provides a framework for the ethical review of research involving humans, it does not fully and completely specify what should be done in key areas, such as risk-benefit analysis and assessment, the selection of research participants, informed consent, remuneration for research participation, compensation for research-related injuries, and the use of ethically tainted data, all of which are discussed in this chapter.[3]

The committee's ethical analysis therefore draws on many different sources in addition to the Common Rule, including the *Belmont Report: Ethical Principles and Guidelines for the Protection of Human Subjects of Research* (National Commission, 1979), Good Clinical Practice (GCP) guidelines (FDA, 2003), the *Declaration of Helsinki* (WMA, 2002), bioethics literature, recent studies by the National Academy of Sciences/Institute of Medicine, a report by the National Bioethics Advisory Commission (NBAC, 2001), a report by the National Human Research Protections Advisory Committee on research with children,[4] and policies and practices

---

[2]Informed consent and review by an Institutional Review Board (IRB) are considered the major pillars in the system for protecting the rights and welfare of participants in research (NBAC, 2001).

[3]This report does not address all the ethical considerations that are relevant to the evaluation of intentional dosing studies. Instead, it concentrates on the ethical considerations that are especially unclear or controversial in intentional dosing studies, while presupposing the other ethical conditions, such as confidentiality.

[4]See ohrp.osophs.dhhs.gov/nhrpac/documents/nhrpac16.pdf.

that have evolved in the conduct of research (see Chapter 2). Even though these sources agree on the importance of Institutional Review Board (IRB) review and informed consent, they are frequently unclear, indeterminate, inconsistent, and even contradictory regarding other issues related to toxicant research. Hence, in this report the committee often presents its own judgments, based on the best available arguments, or recommends a process that over time can more fully address such issues.

Studies in which people are intentionally exposed to toxicants, which are conducted generally to make the case for setting a less stringent exposure standard, are intuitively troubling and even repugnant to many people. Such studies seem to be ethically wrong—"It's wrong to poison people"—and further discussion does not even seem necessary. The committee took note of these responses but sought to examine closely how toxicant studies are similar to, and different from, other human studies, so that the wide experience could contribute to its deliberation about which kinds of studies are ethically defensible in light of the available evidence and society's basic moral values. Understanding that virtually all chemicals can be poisonous to humans at some dose, the committee compared studies that involve the intentional exposure of humans to toxicants with studies that involve deliberate exposure to other kinds of chemicals. This analysis noted some important similarities, along with several differences, between intentional human dosing studies and Phase 1 pharmaceutical testing, especially because neither offers a reasonable prospect of direct benefit to the research participant. In fact, the Phase 1 study is more likely to provoke adverse effects. Both types of study should be evaluated according to prevailing ethical standards, in the Common Rule and elsewhere, for assessing human research protocols. Neither kind of study can be ethically justified unless it passes rigorous scrutiny on both scientific and ethical grounds.[5]

Recommendation 5-1: Criteria for Scientific and Ethical Acceptability

Studies that do not meet the highest scientific and ethical standards should not be carried out or accepted by EPA as input to the regula-

[5]The motives of the different sponsors also may be similar: both a pharmaceutical company and a pesticide company, to take these two examples, may be motivated primarily by a desire for increased revenues. One seeks to get a drug approved for sale, and the other seeks a higher tolerance level to increase the sale of pesticides. These motives may be primary or secondary and may be accompanied by various additional motives. In any case, neither motive necessarily disqualifies the research on ethical grounds. However, the presence and perhaps primacy of these motives underscores the need for stringent standards and procedures to protect the rights and welfare of research participants.

tory decision-making process. Necessary conditions for scientifically and ethically acceptable intentional human dosing studies include:

a. prior animal studies and, if available, human observational studies;
b. a demonstrated need for the knowledge to be obtained from intentional human dosing studies;
c. justification and documentation of a research design and statistical analysis that are adequate to address an important scientific or policy question, including adequate power to detect appropriate effects;
d. an acceptable balance of risks and benefits and minimization of risks to participants;
e. equitable selection of participants;
f. free and informed consent of participants; and
g. review by an appropriately constituted IRB or its foreign equivalent.

Examples of unethical studies include the following:

- studies that are unnecessary because the desired information can be obtained by other means, such as animal studies or human observational studies, without resorting to the intentional exposure of research participants to toxicants;
- studies that lack prior and appropriate animal studies;
- studies that are not designed to yield scientifically valid information that addresses important scientific or policy questions;
- studies that have an unacceptable risk-benefit balance or that fail to minimize risks to participants;
- studies whose selection of research participants is inequitable;
- studies for which the consent of the participants is not informed and voluntary; and
- studies that have not been reviewed by an appropriately constituted IRB.

Other recommendations in this chapter specify these ethical criteria in more detail. Several of them reflect the ethical principles presented in the *Belmont Report*: beneficence, justice, and respect for persons (National Commission, 1979). While the discussion of risk-benefit analysis and scientific validity in the two preceding chapters largely reflected ethical considerations based on the principle of beneficence, this chapter focuses mainly on ethical considerations based on the principles of justice and respect for persons. The principle of justice guided the committee's judgments about the selection and recruitment of participants in research and compensation for research-related injuries, while the principle of respect for persons shaped the committee's recommendations about voluntary informed consent by potential research participants. Both principles are involved in judgments about remuneration for participation in research involving toxicants.

SELECTION OF RESEARCH PARTICIPANTS

According to the Common Rule, IRBs should not approve a research protocol involving humans unless "selection of subjects is equitable" (40 CFR 26.111(3)). This requirement derives from the principle of justice identified in the *Belmont Report*. If a research protocol has a satisfactory overall ratio of risks and potential benefits, it satisfies the demands of beneficence, but not necessarily the demands of justice. The principle of justice directs attention to the *distribution* of risks and benefits—who will gain the benefits and who will bear the risks and other burdens of research—not just the overall risk-benefit ratio (Beauchamp and Childress, 2001; EPA, 2000; National Commission, 1979). It is easier to identify and avoid some unjust distributional patterns—for example, the deliberate selection of certain relatively powerless groups to bear the burdens of research—than it is to design and implement a fully just distribution. Furthermore, as the *Belmont Report* noted, researchers and institutions may lack the power to counteract some social factors, such as socioeconomic status, that result in higher rates of enrollment by members of certain groups (National Commission, 1979).

Several aspects of the ethical requirement of equitable, fair, or just selection of research participants merit attention. First, on scientific grounds, as discussed in Chapter 3, the study population needs to be representative of the target population of interest in order for the research results to be applicable. Although variations in gender and race/ethnicity may not signify the true scope of biologic variation affecting response to toxicants, they may be helpful proxies and thus should be considered in participant selection.

The selection of research participants also should be inclusive in order to avoid the exploitation and the appearance of exploitation of any particular social group. As the *Belmont Report* observed in discussing the principle of justice:

> [T]he selection of research subjects needs to be scrutinized in order to determine whether some classes (e.g., welfare patients, particular racial and ethnic minorities, or persons confined to institutions) are systematically selected simply because of their easy availability, their compro-

mised position, or their manipulability, rather than for reasons directly related to the problem being studied (National Commission, 1979, 7).

The *Belmont Report* and the Common Rule both note that various conditions can render some persons more "vulnerable to coercion or undue influence" and hence create the need for "additional safeguards" to protect their rights and welfare as potential research participants. Potentially vulnerable populations include children, prisoners, persons with mental disabilities, and economically or educationally disadvantaged persons. Vulnerability may reflect limited abilities to make informed choices (e.g., limited mental capacity) or constraints on free choices (e.g., imprisonment or economic disadvantage). From the standpoint of just or fair selection and recruitment of research participants, it is not justifiable to enroll persons who lack the capacity to consent to their involvement, even if surrogate decision makers grant permission, when the research offers them no prospect of direct personal benefit and carries more than minimal risk or when the needed information could be obtained through studies with individuals who have the capacity to consent.

Concerns about voluntariness led the Department of Health and Human Services (DHHS) Office for Human Research Protections (OHRP) to raise questions about "undue influence" when employees and students of institutions sponsoring or conducting the research serve as participants in research that offers no prospect of direct benefit. Concerns about undue influence also may arise in other cases, such as including in studies the employees of companies that make the products being tested. Under some circumstances, a person may feel compelled to participate in a dosing study, perhaps especially when the person or entity conducting the study has substantial power over the potential participant. The Common Rule permits consent to be sought only "under circumstances . . . that minimize the possibility of coercion or undue influence."

A separate issue arises from proposals to enroll individuals who are more susceptible to harm in research protocols, as can occur, for example, in studies of the effect of aerosolized pollutants on individuals with lung diseases such as asthma. As discussed in previous chapters, investigators and IRBs have a responsibility to minimize risks to research participants. Among the several ways available to minimize risk, investigators in intentional human dosing studies usually can select participants without known health conditions that put them at increased risk for adverse effects from the experiment. In general, individuals who would face higher risks in the experiment should not be selected. An exception might be warranted if their participation is needed to answer a question of major importance in the regulatory process. However, even then, this exception should be made only when additional measures are taken to ensure an acceptable balance of risks and potential benefits. These measures could include making sure that a review is conducted of a volunteer's possible participation by his or her physician or another physician not involved in the study, monitoring during the study, and reinforcing the usual medical advice given to patients. Such measures can help to provide an acceptable balance of risks and potential benefits for the individual participant.

Children represent a special case. They are vulnerable because they lack decision-making capacity and are greatly influenced by adults. There also is reason to believe that children may be more susceptible to certain adverse effects of toxicants because of changes that occur during development and because of age-dependent differences in metabolism, disposition, and target organ sensitivity. Infants and toddlers are often particularly susceptible. For example, lead is more toxic to the developing child than it is to adults in both the short and long term. The fear of greater adverse effects is reflected in the requirement of the Food Quality Protection Act that EPA add a 10-fold safety factor to account for children's possible increased susceptibility that can be rebutted only by "reliable data." A major question, then, is whether and, if so, under what conditions, it is permissible to conduct research to learn more about the susceptibility of children.

DHHS has addressed the tension between the need for greater knowledge about children and the need to protect children from harm and exploitation in research. Subpart D of the Common Rule (Additional DHHS Protections for Children Involved as Subjects in Research) greatly restricts the enrollment of children in research that involves greater than minimal risk without the prospect of direct medical or health benefit. Such research may be approved by an IRB if it is likely to yield generalizable knowledge about the children's "disorder or condition" and ways to ameliorate that "disorder or condition," but only if the risk represents "a minor increase over minimal risk," the intervention or procedure "presents experiences to subjects that are reasonably commensurate with . . . their actual or expected medical . . . situations," and the parents grant permission and the children assent. Research that would not otherwise be approvable under these criteria, however, could be approved if the DHHS Secretary, in consultation with a panel of experts, determines that it "presents a reasonable opportunity to further the understanding, prevention, or alleviation of a serious problem affecting the health or welfare of children," the study will be conducted in accord with "sound ethical principles," and the parents grant permission and the children assent (45 CFR 46.407(a)).

EPA should adopt Subpart D, and, in any event, should adhere to its requirements. The provisions of Subpart D leave open the possibility of research involving deliberate exposure of children to toxicants as long as the research undergoes rigorous scrutiny, at times by a nationally constituted panel, and the investigation will increase the understanding of a serious problem affecting the health of children. Simply improving the accuracy of risk assessments for regulatory decision making would not justify research under this subpart.

The ethical problems of conducting dosing studies in children emphasize the importance of conducting rigorous epidemiological studies in children. Nonetheless, if EPA followed the model provided by Subpart D and meets the requirements of that subpart outlined above, then research involving children that otherwise would not be approved could be considered and perhaps approved by a special panel.

Recommendation 5-2: Participant Selection Criteria

IRBs reviewing intentional human dosing studies should ensure that the following conditions are met in selecting research participants:

a. Selection should be equitable.
b. Selection of persons from vulnerable populations must be convincingly justified in the protocol, which also must justify the measures to be taken to protect those participants.
c. Selection of individuals with conditions that put them at increased risk for adverse effects in such studies must be convincingly justified in the protocol, which also must justify the measures that investigators will use to decrease the risks to those participants to an acceptable level.

EPA should adopt Subpart D of the Regulations for the Protection of Human Research Subjects. At a minimum, EPA should adhere to Subpart D's requirements for research involving children.

PAYMENT FOR PARTICIPATION IN RESEARCH

Another issue of justice, as well as of respect for persons, involves remuneration for participation in research.[6] Paying research participants is "a common and long-standing practice in the United States" (Dickert et al., 2002, 368), perhaps because of the need to provide incentives as part of recruitment and because the moral principles of fairness and gratitude

[6]The *IRB Guidebook* (available at ohrp.osophs.dhhs.gov/irb/irb_guidebook.htm) proposes that the term "remuneration" be used for payment for participation in research and that "compensation" be reserved for payment or provision of medical care for research-related injuries.

support providing payment to those who bear the burdens of research on behalf of society. In any event, difficult questions remain: How much money should research participants receive? And for what should they receive payment—their time, inconvenience, discomfort, or level of risk? Can remuneration—or some level of remuneration—create a problem for research subjects' voluntary, informed consent?

Although the consensus is that remuneration for participation in research should be just or fair, there is little agreement in theory or in practice about what constitutes just or fair payment. Disagreement can certainly be expected about payment for participation in intentional human dosing studies. Furthermore, federal regulations and guidance are relatively quiet on this subject, warning about "undue influence" without, however, specifying what counts as undue. One difficulty is that undue influence depends on context. Wherever the remuneration is set, it will influence the decisions of some more than others. In particular, it will be more important to those for whom it will make a significant financial difference, i.e., poor people. Although the committee does not purport to be able to resolve the ethical difficulties surrounding remuneration of research participants, it believes that some general guidance may be useful.

A major ethical concern is that payments should not be so high that they create an "undue influence" or offer undue inducement that could compromise a prospective participant's examination and evaluation of the risks or the voluntariness of his or her choices. This concern is greatest, of course, when the studies involve significant risks. Other concerns are that payments should not be so low as to recruit disproportionately high numbers of economically disadvantaged persons and that they should fairly pay participants for their contribution to research.[7]

[7]One committee member (Lave) believed that this discussion about payment misses the point. Undue influence is important, but the amount of money that constitutes undue influence differs among individuals. An impoverished person might be willing to take considerable risks and bear considerable pain in exchange for a few hundred dollars. Thus, the investigator and the IRB should assume that any reasonable level of payment will unduly influence some potential participants. The experimental protocol must strive to detect cases where a potential participant has concealed information or lied in order to participate, despite the fact that this person has characteristics that greatly increase his risk of being harmed in the experiment, or if he believes that the existence of symptoms would exclude him from the experiment or force his early exit, reducing the amount of his payment. Assuming that all potential participants will tell the truth and conceal no information because the payment does not constitute undue influence is a mistake. Furthermore, participants will deliberately conceal adverse symptoms that would remove them from the experiment, if they believe that doing so would erode the amount of their payment. The protocol must make every effort to get participants to be truthful in revealing their symptoms without fear that remu-

In its guidance on "Payment to Research Subjects," the Food and Drug Administration (FDA) notes that:

> Financial incentives are often used when health benefits to subjects are remote or non-existent. The amount and schedule of all payments should be presented to the IRB at the time of initial review. The IRB should review both the amount of payment and the proposed method and timing of disbursement to assure that neither is coercive or present undue influence (21 CFR 50.20).

In particular, the FDA guidance indicates that payment should be prorated for the time of participation in the study rather than extended to study completion, because the latter could compromise the participant's right to withdraw at any time.

Despite such guidance, there appears to be wide variation among IRBs, sponsors, and investigators in policies and practices involving remuneration for research participation. Indeed, a recent study of institutional policies reports that "few data are available on guidelines used by research organizations to make decisions about paying subjects" (Dickert et al., 2002, 368). This study notes that few of the 32 research organizations surveyed had formal policies to guide the amount or circumstances of payment, a situation that generates uncertainty about whether safeguards against unfair or coercive payment are adequate (Dickert et al., 2002).

According to this study, participants in some research receive payment for time (87 percent of organizations), inconvenience (84 percent), travel (68 percent), incentive (58 percent), or incurring risk (32 percent) (Dickert et al., 2002). In line with these patterns, many argue that research participants should be paid for their time and inconvenience, as well as their expenses, but are concerned about providing payment for incurring risk, which some would rule out altogether. However, attitudes may differ considerably when the risk is a minor and transient symptom or discomfort (e.g., sleepiness or dizziness) rather than a substantial harm. Sometimes the arguments for limiting payment to time and inconvenience reflect the belief that participation in research is an altruistic act. It is al-

---

neration will be eroded. A final point is that a much higher level of remuneration will be required to get potential participants with higher income to volunteer for the experiment. If we take seriously the notion that participants should be representative of the population to be served, remuneration will need to be raised. Higher remuneration means that still more participants are likely to attempt to hide risk factors or lie so that they can participate in the experiment and earn the remuneration. Thus, the attention paid to remuneration is largely wasted, because the protocol needs to focus on detecting factors that would put the participant at higher risk, despite his or her attempts to conceal them or to lie.

most certainly true that the prospect of financial remuneration motivates many people to participate in research and that it is often a necessary and sometimes a sufficient condition for their participation. Indeed, it is difficult to believe that people who agree to participate in dosing studies of toxicants are motivated only by altruism and not by the desire to make money. It thus seems highly likely that remuneration will affect participation.

Because at present there is no practical or theoretical consensus regarding remuneration, the committee recommends that sponsors, investigators, and IRBs closely attend to the ethical and scientific implications of different strategies, particularly regarding payment for incurring risk. Protocols submitted to IRBs should indicate and justify proposed levels and purposes of remuneration, which also should be clearly stated in the accompanying consent forms.

Recommendation 5-3: Payment for Participation

IRBs, all relevant review boards, investigators, and research sponsors should ensure that payments to participants in intentional human dosing studies are neither so high as to constitute undue inducement nor so low as to be attractive only to individuals who are socioeconomically disadvantaged. Proposed levels of and purposes for remuneration (e.g., time, inconvenience, and risk) should be scrutinized in light of the principles of justice and respect for persons.

Moreover, EPA, in conjunction with other federal agencies, should consider developing further guidance on remuneration for participation in intentional human dosing studies, including guidance regarding whether remuneration should reflect the level of risk as well as the time and inconvenience involved.

INFORMED CONSENT

Voluntary, informed consent by research participants (or permission by their surrogate decision makers), is a major element in the system of protection of research participants. The consent requirement expresses the principle of respect for persons, including respect for and promotion of autonomous choices. The Common Rule stresses this requirement, as do other codes of research ethics, including the Nuremberg Code (1949), the *Declaration of Helsinki*, and GCP guidelines. This section focuses on the disclosure and comprehension of information as part of the consent process.

Despite the strong consensus about the importance of informed consent, various studies indicate that those who have agreed to participate in research often do not comprehend its basic features (Joffe et al., 2001; Miller et al., 1996). Although these studies are frequently limited because they focus on what participants later recall, they raise legitimate ethical concerns about the validity of consent in many experiments. Problems may arise, in part, because much ethical discourse focuses on the obligation to disclose information, rather than on the obligation to ensure participant comprehension. For example, the Common Rule specifies much of what should be disclosed to participants, including the nature and purposes of the research, the procedures used, "any reasonably foreseeable risks or discomforts to the subject," and any potential benefit, but it provides less guidance about how to ensure participant comprehension.

The focus on disclosure also results in part from IRBs' attention to the consent *form* rather than to the *process* of consent. Whatever the forms say, and however clearly they say it, incomplete understanding or misunderstanding is common. Even if the consent forms are clear about the experimental nature of the study, a "therapeutic misconception," that is, a mistaken belief that the research offers a real hope of medical or health benefit to participants, may emerge. For example, recruiting advertisements can affect participants' understanding (and copies of these advertisements could be requested by IRBs along with the consent form). Thus, focusing solely on disclosure is not enough; it is appropriate to be concerned about what participants in intentional human dosing studies comprehend. One simple approach would be to administer, at the time of the consent, a short multiple-choice test, which could indicate how well the participants understand the disclosed information (EPA, 2000, 20).

Among the protocols reviewed by the committee, some proposed informed consent procedures that used best practices, while others presented deficient informed consent procedures. Several studies provided information about the research in ways that were ethically problematic:

- A study of toluene inhalation failed to disclose that its purpose was to determine the toxicity of toluene in order to determine exposure levels. The consent form did not say that toluene is an environmental pollutant.
- A study administering particulate air pollutant to normal volunteers involved bronchoscopy as one of the research procedures. The consent form did not list death as a serious but remote risk of this study. In contrast, the low risk of death is routinely disclosed to patients during the consent process for bronchoscopy as a clinical procedure. Furthermore, in one highly publicized study, a healthy research volunteer died after bronchoscopy (Steinbrook, 2002).
- Other studies administered concentrated ambient air particles to

persons with asthma or chronic obstructive lung disease who had been instructed to forgo their usual maintenance therapy. The risks of these studies were understated in the protocol and the consent form. Although the consent forms characterized the research risks as minimal, these studies do not meet the criteria for minimal risk as defined in the Common Rule. While everyday experience can include exposure to comparable levels of such particles, it is not usual for persons with asthma or chronic obstructive lung disease to withhold their customary bronchodilator or to exercise strenuously under such circumstances. Furthermore, the consent forms did not clearly state that there was risk the study could induce an asthma attack or exacerbate chronic obstructive lung disease, which could be avoided if the participant continued to take his or her regular asthma medications or did not participate in the study at all. Finally, it is misleading to compare the risk of the research to the risk of visiting a large (polluted) city, because those with asthma would be advised to take their regular medications when visiting a polluted city—or even to increase them—and to stay indoors to avoid exposure to pollutants.[8]

In light of the documented problems with informed consent and its importance in helping to assure the ethical integrity of intentional dosing studies, the committee recommends that steps be taken to strengthen the informed consent process. One way to do this is to identify best practices regarding informed consent in such studies and to encourage other investigators to adopt them.

Recommendation 5-4: Best Practices in Informed Consent

EPA should develop and disseminate to relevant IRBs, investigators, and sponsors a list of best practices regarding informed consent in intentional human dosing studies. EPA should encourage all sponsors and investigators to adopt these practices, and it should require their adoption in studies it sponsors or conducts.

The initial version of this best practices document should include but not be limited to the following:

Practices to describe the purpose of the study clearly in laypersons' terms. Some studies convey in clear, simple lay language how toxicant studies differ from clinical trials:

[8]These three studies and the associated consent forms were approved by IRBs. Two were approved by research university IRBs and one by an industry-sponsored IRB.

- "The purpose of this research study is to determine what dose of a pesticide can be safely administered to human beings."
- "This study is not designed to provide you a direct medical benefit."
- "This study is not designed to improve your health."

Practices to describe the risks clearly in laypersons' terms, including remote but serious risks. Toxicant studies need to make clear to potential participants the risks of the study, particularly those that are unlikely to be obvious:

- Risks should include any requirements to stop usual medications or deviate from usual medical advice.
- The remote risk of death to a healthy volunteer from an invasive procedure such as bronchoscopy needs to be described. This is standard practice in clinical medicine, where such procedures are performed with the prospect of direct clinical benefit to the person undergoing the procedure. In a research setting that is not designed to provide benefits to participants, the level of disclosure of risks should be even greater than in clinical practice.
- Potential participants should be informed of any reproductive risks or risks to offspring.

Practices to assess whether participants comprehend the information disclosed to them. Disclosure of information to prospective participants in research is only one step in the consent process. It also is essential that the participants comprehend the disclosed information and how it applies to their decision to enter the study. Concerns have been raised that participants may not understand how toxicant research differs from clinical trials that hold the prospect of direct clinical benefit from the administered substance. To allay these concerns, it is important to ensure that participants in intentional dosing studies understand crucial information that has been disclosed, and this generally will require direct assessment of comprehension. Researchers can learn from other studies how best to carry out these assessments.

Such a list of best practices should be used to stimulate investigators and IRBs to consider what consent procedures would be most appropriate for a particular study, not as inflexible requirements that must be applied in every case. A practice that works well in one study may not be appropriate in another. In addition, a list of best practices is not meant to be exclusive. A research team may devise an approach that constitutes an innovative advance over previous consent procedures. The goal of the list is to focus attention on the consent process and to encourage investigators and IRB members to consider how to strengthen it.

COMPENSATION FOR RESEARCH-RELATED INJURIES

Debate continues in the United States about whether compensation should be provided for research-related injuries. The Common Rule requires only that when research involves more than minimal risk, information should be disclosed regarding whether medical treatment and other compensation will be provided for research-related injuries. Many critics of the U.S. policy believe there should be more than disclosure of information about compensation and call for the provision of medical care for research-related injuries without cost to the participants and, in addition, for compensation for lost wages, disabilities, and death. These claims are based on the belief that research participants, whatever their motivations, accept risk on behalf of society. When participants are injured, justice, fairness, and gratitude mandate, at a minimum, the provision of needed medical treatment without cost to the participant. Further study is needed regarding the provision of other types of compensation.

In the United Kingdom, several studies involving deliberate exposure to toxicants indicated that participants who were injured in the research would receive compensation. For example, a Dichlorvos study conducted by Medeval, a malathion study conducted by Cheminova, and a Phosmet study conducted by Inveresk provided for no-fault compensation for physical injuries caused by the research. One study stated that in the event of "any bodily injury caused by my participation in the study," compensation would be paid "without having to prove that the injury arose through negligence or that the study compound was defective." The amount of compensation "shall be calculated by reference to the amount of damages commonly awarded for similar injury by an English court if liability is admitted." In addition, some of the studies in the United Kingdom provided for monetary compensation if long-term disability resulted from injuries incurred during the studies.

In countries with universal access to medical care, research participants would be expected to receive medical care for injuries suffered in research regardless of the cause. That cannot be assumed in the United States. The committee concludes that justice and fairness require sponsors of intentional human dosing studies to go beyond existing legal requirements and create a mechanism that, at a minimum, ensures that research participants receive free medical care for injuries incurred in the research. As NBAC writes:

> Because the costs of research injuries should not be borne by the injured participants and because support for a compensation system should be provided by those most likely to profit or derive other benefits from it, sponsors and institutions should be assigned responsibility for funding such a system (2001).

The operation and scope of such a compensation system require further attention, because of difficulties in determining causation when medical problems appear some time after participation in research.[9]

Recommendation 5-5: Compensation for Research-Related Injuries

At a minimum, sponsors of or institutions conducting intentional human dosing studies should ensure that participants receive needed medical care for injuries incurred in the study, without cost to the participants.

In addition, EPA should study whether broader compensation for research-related injuries should be required.

THE USE OF RESULTS FROM ETHICALLY PROBLEMATIC STUDIES

A final question concerns what role, if any, ethically problematic or unethical studies should play in EPA's regulatory decisions. The committee concludes that this question will rarely arise, especially after EPA formulates its standards and procedures, and worries that it may be magnified out of proportion to its overall frequency. Nonetheless, when this question does arise in real cases, it can be an ethically vexing one. In addressing it, the committee considered the relevance of several distinctions: those between data submitted by industry as part of EPA's process of regulatory decision making and data retrieved by EPA; between data drawn from studies conducted before EPA's anticipated rulemaking in light of this committee's recommendations and studies conducted afterwards; and between the failure to obtain voluntary informed consent and the failure to realize other ethical standards. The committee concludes that, as a general rule, EPA should not use data from ethically problematic studies to inform its regulatory efforts.

Studies Submitted for Regulatory Decisions After EPA Establishes New Standards

After EPA establishes new rules and procedures, those who submit data from intentional human dosing studies should produce evidence that

[9]The report of the EPA SAB and the FIFRA SAP held that participants in research "should have rights to compensation if they are injured as a result of the experiment" (EPA, 2000, 21). The report notes that, because injuries may not become evident until long after the study has ended, investigators need to indicate to IRBs (and in consent forms) their "plans for ascertaining the subjects' health status for some period after the end of the experiment, and ensure that each subject is given clear information about how to deal with problems that might emerge later"(21).

those studies were conducted ethically and in accordance with the new rules and procedures. After the proposed new procedures have been fully implemented, those who submit studies will presumably have had the benefit of advance protocol review by the EPA Human Studies Review Board proposed in Chapter 6, as well as EPA's clarification of the relevant ethical standards.

It also will be necessary for the EPA Human Studies Review Board to review submitted studies in order to determine whether they were ethically conducted. If the research is determined to be unethical, two important goals may come into conflict: first, using the best scientific data to protect the public and, second, avoiding incentives for the conduct of unethical research involving humans and undermining important ethical principles.

If the EPA Human Studies Review Board determines that the submitted research breached fundamental ethical standards, but also determines that the data would be important in protecting the public, what should it do? In such cases, the committee recommends that EPA convene a special, outside panel (distinct from the Human Studies Review Board) to examine the case for and against using the data. Such an exceptional procedure signifies the seriousness of any possible reliance on data from research that violates important ethical standards. The outside panel should include members of the public as well as experts, because the judgments that are required are not only scientific and technical but also involve societal values and because the judgments will need to be explained and justified to the public. Even though the panel will need to specify the relevant substantive standards as it wrestles with real cases, the following points are relevant.

The panel should first determine whether the data are "crucially important" for protecting public health and whether they are necessary in the sense that they could not otherwise be obtained, with reasonable certainty within a reasonable period, without exposing additional research participants to harm. In part because this standard's key terms and concepts are imprecise, the panel's judgment will be required in determining whether the answers to both questions are affirmative.

It is critically important for EPA to deter future unethical conduct even as, in extraordinary circumstances, it considers and relies on data from unethical research to protect public health. In such circumstances, the committee concludes that it would be possible, through the creation of the special panel described above, and through adherence to stringent substantive standards, to use unethically obtained data to protect the public without creating an incentive for future breaches of the relevant ethical rules.

Nonetheless, some argue that it is not sufficient to establish safeguards to prevent future ethical abuses; instead, they contend, EPA should totally reject ethically tainted research data. This argument charges that deriving societal benefits from unethical research retrospectively legitimizes such research and undermines the ethical principles discussed earlier in this chapter. Indeed, this argument holds that accepting the benefits of such research involves society in a kind of symbolic approval of and complicity in the unethical research, even after the fact. This line of reasoning tends to support an absolute renunciation of the benefits of knowledge gained through unethical research.

Although this stance has strong appeal, especially as a way to express society's commitment to fundamental values in research involving humans, it would sacrifice another important societal value, namely, the protection of public health. It is difficult enough to resolve this debate in concrete cases—as was evident in the dispute several years ago about whether EPA should use data from Nazi experiments on the effects of phosgene. However, it is virtually impossible to resolve this debate in the abstract, especially when the kinds of cases envisioned are not as egregiously or as blatantly unethical as the Nazi experiments, which included the intention to harm research subjects. Thus, instead of attempting to resolve this dispute in the abstract, the committee recommends the conduct of a rigorous review by the special, outside panel of actual cases using stringent substantive standards that should, at a minimum, prevent the creation of incentives for any future abuses.

Recommendation 5-6: Studies Completed After Implementation of the New Standards

EPA should operate on the strong presumption that data obtained in studies conducted *after* implementation of the new rules[1] that do not meet the ethical standards described in this report will not be considered in its regulatory decisions. Under exceptional circumstances, studies that fail to meet these ethical standards may provide valid information to support a regulatory standard that would provide greater protection for public health. Under these circumstances, EPA should convene a special, outside panel, consisting of relevant experts and members of the public, to examine the cases for and against considering data from such studies.

Enacting regulatory standards based on data from such studies without requiring toxicants to be administered to additional people to repli-

[1]The committee uses the term "rules" informally to mean guidance, guidelines, policy, protocols, rules, or regulations.

cate them might better protect the public health, but in order to strongly deter sponsors and researchers from conducting unethical studies, these data should not be used to favor the sponsor's interests in loosening regulatory standards.

The special outside panel (convened by EPA or as an ad hoc panel of the Human Studies Review Board) should make its judgment by considering:

(1) whether the data are crucially important for protecting the public, and

(2) whether the data cannot otherwise be obtained, with reasonable certainty within a reasonable period, without exposing additional research participants to the risk of harm.

Unless the panel can answer both questions affirmatively, it should recommend that EPA not consider the data in question.

Studies Completed Before Publication of EPA's New Rules

Consideration of the use of data that were collected before the new standards are in effect raises particularly vexing issues. One question is whether it is fair to judge past studies with humans by current ethical standards. To be sure, some ethical standards proposed in this report for future intentional human dosing studies have only been articulated or at least stressed in recent years (e.g., just selection of and fair payment to research participants), and some remain unsettled (e.g., compensation for research-related injuries). However, informed consent has earlier roots, for instance, in the Nuremberg Code's emphasis on voluntary consent in the late 1940s and in the *Declaration of Helsinki's* attention to informed consent from the 1950s on. And IRB review has been considered an important procedure for ethical review since it was required in 1966 for human research funded by the Public Health Service.

The options range between the following two basic policies:

(1) Reject all studies that do not provide clear evidence that they meet standards for ethical research involving humans.

(2) Accept all studies unless they violated fundamental ethical standards.

The evidentiary requirements for these two options differ. In the first, the researchers must provide the evidence of compliance with ethical standards; in the second, EPA would accept the studies unless there is evidence of the violation of fundamental ethical standards. The committee favors the second option because of ethical concerns about not considering scientifically valid data from completed studies. If such data are not considered, it may be necessary to conduct additional research to obtain similar data to protect the public, thus subjecting additional research participants to risk.

Moreover, it would be difficult and often impossible to obtain evidence about whether past studies, especially those in the distant past, met ethical standards. Adequate documentation often is not available. Publications before 1975 do not usually indicate whether investigators obtained IRB approval and informed consent. This lack of documentation is true even for federally funded studies, which, after 1966, were required to obtain IRB approval and informed consent. In some medical specialties, even more recent publications do not consistently state whether informed consent and IRB approval were obtained, and even when publications do mention informed consent and IRB review, they almost never provide the kind of detailed information that would be required by the Human Studies Review Board in its review. Furthermore, for older studies, it may be difficult to obtain copies of the protocol or consent forms and procedures if the investigator has retired or died.

Recommendation 5-7: Studies Completed Before Implementation of EPA's New Standards

EPA should accept scientifically valid studies conducted before its new rules[2] are implemented unless there is clear and convincing evidence that the conduct of those studies was fundamentally unethical (e.g., the studies were intended to seriously harm participants or failed to obtain informed consent) or that the conduct was deficient relative to then-prevailing ethical standards. Exceptional cases in which the Human Studies Review Board determines that unethically conducted studies may provide valid information to support a regulatory standard that would provide greater protection for public health should be presented to a special, outside panel, described in Recommendation 5-6, for consideration.

This special, outside panel should consider recommending the use of such data only with the requirement that the ethical concerns raised by the study are documented and made publicly available, along with relevant materials and commentary, on the EPA web site.

Recommendation 5-7 applies both to studies submitted to EPA as part of the regulatory process and to studies that EPA has retrieved from the

[2]See footnote 1.

literature. More specific questions have arisen about a number of third-party studies that were completed and submitted to EPA after the mid-1990s. There is debate about whether EPA should consider and rely on these studies of deliberate exposure to pesticides. According to the committee's recommendation, EPA may consider and rely on them if they provide scientifically valid and relevant data, unless there is evidence that they violated fundamental ethical standards or the then-prevailing ethical standards. Because these studies were conducted with a view to submission of the data to EPA as part of its regulatory decision making, more evidence should be available about their compliance with certain ethical standards governing research involving humans. Specifically, for such recent studies, it would be expected that the full protocol, consent forms and procedures, and documentation of IRB approval would be available. After all, EPA's long-time standards already exclude certain unethical conduct of the sort envisioned by this recommendation. One such standard appears in the October 21, 1972, amendments to the Federal Insecticide, Fungicide, and Rodenticide Act (P.L. 92-516). According to this statute, it is unlawful to test pesticides on humans unless they are fully informed about the tests' nature and purposes as well as any reasonably foreseen health effects and they freely volunteer to participate (EPA, 2000, 30; Latham and Watkins, 2003, 2).

In the public comments on the EPA notice of proposed rulemaking, pro-industry advocates argued that it would be unfair and illegal to hold studies to standards that were not legally required at the time the study was conducted. However, ethical guidelines may be morally binding even if they are not legally binding. Some ethical lapses—such as the intention to seriously harm participants—violate universal and timeless ethical principles even if they are technically not legally prohibited. Similarly, carrying out an experiment without the permission of participants or of their surrogates would be considered a grave ethical failure. Even if there were no legally binding requirement for informed consent from participants, the Nuremberg Code of 1949 and the *Declaration of Helsinki* of 1964 clearly establish that failure to obtain informed consent from participants is unethical, and a requirement for consent was included in the Kefauver-Harris Amendments to the Food, Drug, and Cosmetic Act in 1962.

SUMMARY

This chapter addressed the ethical considerations that remain after the determination is made that a research protocol is scientifically valid and that its probable benefits outweigh its risks. These ethical considerations involve voluntary informed consent and the fair selection and recruitment of potential research participants, including fair payment for participation and compensation for research-related injuries (which includes the provision of medical care without cost to participants injured in research). After analyzing how these ethical considerations apply to toxicant studies, the chapter examined the arguments about whether EPA may use data from ethically problematic and unethical studies for regulatory purposes. The committee concludes that, as a general rule, EPA should not use data from unethical studies. However, the committee also recommends standards and procedures for exceptional cases in which information from such studies would support a regulatory standard that provides greater protection for public health.

REFERENCES

Beauchamp, T. L., and J. F. Childress. 2001. The Principles of Biomedical Ethics. New York: Oxford University Press.

Dickert, N., E. Emanuel, and C. Grady. 2002. Paying research subjects: an analysis of current policies. *Annals of Internal Medicine* 136(5):368-373.

Environmental Protection Agency (EPA). 2000. *Comments on the Use of Data from the Testing of Human Subjects: A Report by the Science Advisory Board and the FIFRA Scientific Advisory Panel.* EPA-SAB-EC-00-017. Washington, D.C.: EPA. Available at www.epa.gov/science1/pdf/ec0017.pdf.

Food and Drug Administration (FDA). 2003. *Good Clinical Practices.* Available at www.fda.gov/oc/gcp.

Institute of Medicine (IOM). 2003. *Responsible Research: A Systems Approach to Protecting Research Participants.* Washington, D.C.: National Academy Press.

Joffe, S., E. F. Cook, P. D. Cleary, and J. C. Weeks. 2001. Quality of informed consent in cancer clinical trials: a cross-sectional survey. *The Lancet* 358:1772-1776.

Latham and Watkins, LLP. 2003. Re: Protection of Human Participants in Third Party Research STLP-Q-02-02-A. Letter to Anne-Marie Mazza, NAS, on behalf of CropLife America, June 20, 2003.

Miller, C. K., D. C. O'Donnell, H. R. Searight, and R. A. Barbarash. 1996. The Deaconess Informed Consent Comprehension Test: an assessment tool for clinical research subjects. *Pharmacotherapy* 16(5):872-878.

National Bioethics Advisory Commission (NBAC). 2001. *Ethical and Policy Issues in Research Involving Human Participants: Vol. 1.* Bethesda, MD: U.S. Government Printing Office.

National Commission for the Protection of Human Subjects of Biomedical and Behavioral Research (National Commission). 1979. *Belmont Report: Ethical Principles and Guidelines for the Protection of Human Subjects of Research.* Washington, D.C.: U.S. Government Printing Office.

Nuremberg Code. 1949. *Trials of War Criminals before the Nuremberg Military Tribunals under Control Council Law No. 10, Vol. 2*, 181-182. Washington, D.C.: U.S. Government Printing Office.

Steinbrook, R. 2002. Improving protection for research subjects. *The New England Journal of Medicine* 346(18):1425-1430.

World Medical Association (WMA). 2002. *Declaration of Helsinki: Ethical Principles for Medical Research Involving Human Subjects* (adopted 18th WMA General Assembly. Helsinki, Finland, June 1964; amended: 29th WMA General Assembly, Tokyo, Japan, October 1975; 35th WMA General Assembly, Venice, Italy, October 1983; 41st WMA General Assembly, Hong Kong, September 1989; 48th WMA General Assembly, Somerset West, Republic of South Africa, October 1996; and 52nd WMA General Assembly, Edinburgh, Scotland, October 2000. Note of Clarification on Paragraph 29 added by the WMA General Assembly, Washington 2002). Ferney-Voltaire, France: WMA.

6

Procedures for EPA Review of Intentional Human Dosing Studies

INTRODUCTION

Intentional human dosing studies require not only strict substantive restrictions, but also careful procedural requirements to guarantee that the substantive restrictions are followed and ethical standards are met. This chapter delineates those procedures, which include the recommendation that all proposed experiments be reviewed by a properly constituted Institutional Review Board (IRB) and by a new body to be constituted at the Environmental Protection Agency (EPA).

INDEPENDENT REVIEW OF INTENTIONAL HUMAN DOSING STUDIES

As described in Chapter 2, in the United States for the last several decades most federally funded human research has been covered by the Common Rule, which requires advance approval of such research by local IRBs. The Common Rule applies only to human research conducted or funded by signatory federal agencies and to any research performed at an institution that has promised to have all research reviewed by an IRB as part of its assurance of compliance with the Common Rule. In addition, the Food and Drug Administration (FDA) requires review of protocols by an IRB and informed consent of participants for any investigations of new drugs conducted in the United States. On the other hand, private sponsors of intentional human dosing studies submitted to EPA are not specifically required to obtain IRB approval for studies, particularly if the studies are conducted at institutions that do not require IRB review of all research. Although it appears that all of the pesticide experiments reviewed by the committee were approved in advance by IRBs or their foreign equivalents, the committee believes that this decision should not be left to the discretion of the sponsors.

Recommendation 6-1: IRB Review of All Studies

EPA should require that *all* human research conducted for regulatory purposes be approved in advance by an appropriately constituted IRB or an acceptable foreign equivalent. Research conducted by EPA scientists should be reviewed by an EPA-authorized IRB.

EPA may wish to use FDA's implementation of its equivalent of the Common Rule (21 CFR Part 50) as a guide for its adoption of such a requirement (see Chapter 2).

EPA itself has sponsored intentional human dosing studies. At least some of those experiments were approved by IRBs at the institutions that conducted the research. The committee was informed that EPA does not have an IRB, but instead has an Ethics Review Officer who typically ensures that EPA-sponsored or conducted studies undergo IRB review.

If all EPA-sponsored human research is conducted at nonfederal institutions and those institutions have appropriate IRBs operating in compliance with the Common Rule, the federal requirements might be satisfied. If EPA were to conduct human research in-house without prior IRB review, it would be violating the Common Rule.

The IRB that reviews EPA-sponsored research should contain the requisite expertise to review human dosing studies (as well as other human research in support of EPA's mandate). In preparing this report, the committee spent considerable time assessing protocols that were submitted to EPA for registration purposes and that were made available to the committee and/or its staff. In reviewing these protocols, the committee was tasked with ascertaining the meaning and applicability, in the context of EPA research, of "procedures . . . which do not unnecessarily expose subjects to risk" (40 CFR 26.111(a) (1)) and "the importance of the knowledge that may reasonably be expected to result" (40 CFR 26.111(a) (2)). Despite the expertise of the toxicologists, clinicians, and biostatisticians on the committee, this was not an easy task, but it was one that led the committee to appreciate the difficulty of making these determinations. Thus, for EPA's IRB to carry out its assigned duties in reviewing intentional human dosing studies, it must include members with the range of disciplines and perspectives and the array of skills needed for this task. This requirement is equally true for IRBs reviewing studies sponsored by or conducted by non-EPA institutions. All IRBs that undertake the review of intentional human dosing studies should have appropriate scientific and ethical expertise, which may require the appointment of additional members or consultants with expertise in toxicology or biostatistics.

CREATION OF AN EPA HUMAN STUDIES REVIEW BOARD

Despite their limitations, IRBs remain a crucial part of the system of protection for participants in research. They do, however, in special situations require substantive supplementation. For example, gene transfer protocols receive not only local IRB review, but are also subjected to close and public scrutiny by the National Institutes of Health's (NIH's) Recombinant DNA Advisory Committee (RAC). This additional review was instituted after the death of a healthy volunteer in a gene transfer study, which raised concerns that local IRBs may lack the expertise needed to review such studies. The RAC includes scientists from various disciplines pertinent to gene transfer, as well as members with expertise in the ethics of human research. As another example, Subpart D of 45 CFR 46 contains a provision for the Secretary of the Department of Health and Human Services to convene a panel of experts to review research with children that is not otherwise approvable, but that presents an opportunity "to understand, prevent, or alleviate a serious problem affecting the health or welfare of children" (45 CFR 46.407).

Similarly centralized and elevated review would be useful for any proposed intentional human dosing studies conducted for EPA regulatory purposes, given their unique risk-benefit calculus, whether EPA sponsored or sponsor initiated. Previous chapters of this report have highlighted the difficult and controversial ethical and scientific issues involved in some of these experiments, particularly those concerning pesticides. The committee concludes that EPA should not consider such experiments unless they meet high scientific and ethical standards (see Chapters 3, 4, and 5).

The committee also concludes that another level of review is required for these studies to establish a body of knowledge and expertise that can then be communicated and disseminated to the pertinent members of the research community. The committee understands that adding additional review burdens research with additional costs, which may not be trivial. The committee concludes that for this kind of research, however, the benefits of such review outweigh the costs, as such review may provide valuable advice to study sponsors regarding how to structure these experiments. In addition, it was not clear to the committee that local IRBs can be expected to conduct a thorough assessment of this kind of research, as evidenced by problems in IRB review of some of the air pollution studies. The envisioned additional review would provide greater specialized expertise than IRBs usually will have available in considering the special scientific and ethical problems raised by this kind of research, as well as in assessing the potential benefits of the research. Because this research is publicly sensitive, additional review will help build public trust that, when approved, the research is appropriate.

Recommendation 6-2: Human Studies Review Board

To ensure that intentional human dosing studies conducted for EPA regulatory purposes meet the highest scientific and ethical standards, EPA should establish a Human Studies Review Board to address in an integrated way the scientific and ethical issues raised by such studies. To the extent possible, this board should review in a timely manner the protocols and the justification for *all* intentional dosing studies intended for submission to EPA, as well as study results when completed. These reviews should be conducted regardless of the sponsor or site of performance, and EPA should communicate the results of the reviews to relevant parties.

The proposed board's basic function would be to help assure that EPA only uses intentional human dosing studies that meet the rigorous scientific and ethical standards specified in Chapters 3 through 5. This new board would undertake an integrated evaluation of the science and ethics of human research studies (IOM, 2003). In its review of study protocols, the proposed board would not function as a national IRB. Instead, it would provide advice to both the sponsors of the research and to the IRBs that would review it. The board would give the sponsors proposing such research (including EPA itself) advice on how to meet the high standards required. Its report would provide to any reviewing IRBs expert analysis that would help them consider such protocols. After the experiments are completed and the results are submitted to EPA, the Human Studies Review Board would advise EPA's relevant program offices on whether, and to what extent, the results should be considered. The board also would collect and analyze information about these experiments that could allow it to suggest ways in which the research could be improved or to assess whether EPA should continue to consider the results of these experiments.

The committee recommends that the board be relatively small and that it should report directly to the Office of the EPA Administrator. The committee considered whether the board's functions could be discharged within EPA's existing structure. In light of the types of expertise that would be needed in both science and ethics, the committee concludes that no existing EPA office could perform the necessary tasks. Either the EPA Science Advisory Board (SAB) or the Federal Insecticide, Fungicide, and

Rodenticide Act Scientific Advisory Panel, with appropriately enhanced ethical and trial design expertise, might be able to perform those tasks; however, EPA would have to determine whether performing these enhanced functions would interfere with the current obligations of those bodies. Finally, and perhaps most importantly, creating a new board accountable directly to the Office of the EPA Administrator would highlight the importance of this new level of review.

With appropriate staff support, the board's work could be performed by a relatively small but broadly knowledgeable group of experts, with core expertise in human toxicology, biostatistics, and research ethics. It should have the ability to make use of expertise on special subject matters as needed, either by expanding the board temporarily or by using other EPA experts as consultants. In addition, it should be encouraged to coordinate its efforts, as appropriate, with EPA's SAB, SAP, its Ethics Review Officer, and EPA's authorized IRB. Creation of the proposed board raises many other questions, such as conflict of interest limitations and compensation for service, which EPA should address.

The committee deliberated at some length over whether the board should be internal or external to EPA. Arguments can be made for either approach. However, on balance, the committee determined that a formal, permanent, internal structure would be best suited for the kind of integrated scientific and ethical review it envisioned. First, there seems no clear alternative to EPA for the location of such a body, and the creation of a body independent of EPA seems both logistically and politically complex. Furthermore, through the creation of an internal body EPA would take responsibility for the structure and own both the process and the results. The board over time would further specify the general ethical principles and conditions for justified research and develop a series of case judgments and commentaries, enhancing its ability to conduct the best possible reviews. Although the board would be an internal one, EPA could and should invite outside individuals to participate in order to ensure that the necessary expertise is included. Importantly, the board would not replace local IRB review. Rather it would supplement local review by looking at the toxicological and ethical aspects of protocols, and it would assist in improving and refining the science required as part of EPA's regulatory mission (see Chapter 7 and its recommendations for EPA review and the use of scientific data).

The committee also strongly recommends that the board report directly to the Office of the EPA Administrator, rather than be located in any one EPA operational unit, for two reasons. First, it should review experiments sponsored by or relevant to many different EPA functions. Much of the testimony to this committee focused on pesticides, but the committee is making broad recommendations encompassing all intentional human dosing studies sponsored by or submitted to EPA, whether they involve pesticides, air pollutants, water pollutants, or any other relevant EPA jurisdiction. Second, placement of the board within an operational unit of EPA would raise the possibility that the unit's interests might conflict with the board's free consideration of scientific and ethical issues.

Pre-Experiment EPA Review

The committee recommends that the Human Studies Review Board should review in advance proposed intentional human dosing studies sponsored or conducted by EPA and all such studies whose sponsors intend to submit them to EPA. The committee concludes that it would be optimal if this review were mandatory, but, because of legal and logistical concerns, it recommends only that EPA consider making it mandatory. Any conclusions reached by the board should be advisory and not binding on the sponsoring companies or reviewing IRBs. The board's process should take place in advance of local IRB review (or EPA internal IRB review for EPA-conducted studies). It would not replace IRB review. The committee also recommends that the results of this review be made public, taking into account the sponsors' need to protect trade secrets.

As detailed earlier in this report, intentional human dosing studies raise many difficult issues about their scientific worth and ethical propriety. The committee concludes that if such experiments promise scientifically valid and important information, they can be conducted ethically. Instituting a process of advance screening and advice should improve the scientific and ethical quality of protocols. It could lead sponsors to abandon some ill-conceived experiments while, in other cases, it would provide advice that will improve both the scientific value and ethical acceptability of the experiments.

The committee envisions a process similar to the one through which FDA often provides informal advice to firms that are conducting clinical trials on a new drug or biologic. Anyone seeking to experiment with drugs or biologics on humans is required to file with FDA an Investigational New Drug application (IND) and to amend the IND with any new protocols. FDA can put any study it considers unsafe on hold (not allowing it to proceed), or it can engage in discussions with the sponsor about improving the proposed trials. FDA also can put studies on hold if they are not likely to provide useful information—certainly FDA has an interest in identifying studies of no scientific value. In addition, FDA is available to discuss approaches more generally, even before protocols are written. These discussions often lead to clinical trials that are more scientifically valuable, safer, and thus more ethically appropriate than they would have been without the FDA-sponsor interaction. However, FDA has no central board, but rather staff and numerous advisory boards that review protocols in various drug and/or device categories.

The committee discussed at length whether pre-experiment review by the proposed board should be mandatory or voluntary. The main argument for mandatory review was the importance of this review process. The committee wants to prevent inappropriate intentional human dosing studies whenever possible, and requiring review of proposed experiments in advance would lead to fewer inappropriate studies. In addition, making pre-experiment review mandatory should build public confidence that problematic experiments are being minimized and would guarantee that EPA knew of all relevant industry-sponsored experiments, making it impossible for sponsors to keep EPA from learning of experiments that yielded negative results.

Some committee members, however, argued for a voluntary system for experiments not sponsored by EPA. (Everyone agreed that all proposed EPA-sponsored intentional human dosing studies should be reviewed in advance by the board.) They pointed out that few, if any, sponsors would refuse an opportunity to obtain early advice from the board, particularly when it also would review the completed experiment. They further noted that a voluntary system could be easily implemented, while a mandatory system would appear to require, at a minimum, changes in EPA regulations, and possibly new legislation. A voluntary system also would avoid an implementation problem inherent in a mandatory system—the need to distinguish between studies *intended* for submission to EPA, for which the pre-experiment review would be mandatory, and studies independent of a commercial sponsor that later turned out to be relevant to an EPA decision. Of course, if experience were to reveal that many protocols were not submitted for advance review, EPA could take steps to require such submission.

Ultimately, the committee concludes that pre-experiment review of studies intended for submission to EPA *should* be mandatory, if legally and logistically feasible. If not, EPA should strongly encourage study sponsors to seek such review.

The committee strongly urges that any research sponsored, funded, or conducted by EPA that intentionally exposes research participants to toxic substances also should be submitted to the Human Studies Review Board, to ensure consistency in EPA evaluations of such studies and educate EPA program offices about the issues involved and the board about such research within EPA. As discussed previously, under the Common Rule research conducted, sponsored, or funded by EPA also must have IRB approval.

In terms of the sequence of submissions to the board and the local IRB, the committee believes it would be beneficial generally to have each proposed third-party study submitted to the board in advance of the local IRB review, because the board would probably have greater scientific expertise in evaluating these experiments than most IRBs. The board would then offer its views first, which the sponsor would be required to forward to the IRB, thus assisting the IRB's evaluation of such studies. The sponsors should have to submit the IRB report to the board, which would serve to provide feedback to the board from the IRB's perspective—the model RAC uses at NIH. Of course, such review must be done in a timely manner.

The board's pre-experiment review would be advisory. It would make nonbinding recommendations to researchers or sponsors of research about the scientific and ethical aspects of protocols. However, it is likely that the advice of the board usually would be accepted, given the role of the same board in reviewing research results later. The committee does not believe, however, that the board would require veto power over this kind of research, in the way that the IND process gives veto power to FDA. Although a sponsor could proceed with an experiment in the face of a negative board conclusion, the committee believes that few, if any, sponsors would do so without compelling arguments to support their position.

Another issue debated by the committee with regard to pre-experiment review was whether results should be made public. On the one hand, the general availability of the reviews would guide other applicants and help to reassure the public that only scientifically valuable and ethically appropriate studies were being conducted. The NIH RAC has adopted this model, and its deliberations, as well as its conclusions, are public. On the other hand, sponsors may well have legitimate concerns about disclosure of trade secrets or other confidential business information. FDA's discussions with IND applicants are not public, either in substance or results. The RAC also does not make trade secrets public. Although RAC members do see details of vector and gene construct, they promise to keep this information confidential. This is a valuable model for allowing reviewers to see all pertinent information while respecting the confidentiality of trade secrets.

The committee decided to seek the best of both systems and recommended that the board's deliberations should not be public, but that reports on its deliberations and conclusions should be publicly available, except to the extent that they might reveal protected trade secrets or confidential business information. Alternatively, the board could hold public sessions and convene in closed session to review confidential materials. For many of the pesticides undergoing registration after decades of use, the committee expects that few, if any, legitimate claims of trade secrets or confidential business information will be of issue. The board should, however, monitor the effectiveness of the recommended compromise solution and may well, based on its experience, choose to make changes.

Post-Experiment EPA Review

When the results of an intentional human dosing study are submitted to EPA for its regulatory consideration, including studies conducted, sponsored, or funded by EPA, those results should be submitted first to the board for its review. The review should be based on all information collected as part of the study and reported with completeness comparable to that required by FDA for clinical trial submissions. In Chapter 7 the committee recommends a process for internal EPA review of scientific data submitted for regulatory decisions. The results of that staff review should be communicated to the Human Studies Review Board for a second level of review of the scientific value and ethical propriety of the experiments. This model of dual review is used at NIH. The Human Studies Review Board would then provide recommendations to EPA on the scientific and ethical acceptability of such studies. The results of the board's review should be made public, subject to legitimate claims of trade secrets or confidential business information. (The board also may need to consider some delays in the publication of some parts of its report to allow the private investigators to publish their results.) The board also should review studies submitted to EPA that were completed before the effective date of the changes proposed by this report and other studies submitted for EPA's consideration, including those submitted as part of the peer-reviewed literature, for the purposes set out in Chapter 5.

The post-experiment review function of the board is distinct from the kind of review that EPA undertakes for the purpose of incorporating results from particular experiments into the regulatory process. It would not replace or modify the structures and procedures for the latter kind of review. Instead, it would offer nonbinding advice to the relevant EPA units about the scientific and ethical acceptability of the submitted and completed research, not about whether the research should alter the standards for human exposure to toxic substances.

The committee considered whether the pre-experiment and post-experiment reviews should be conducted by the same body. It determined that consistency in judgments and the need for the companies conducting or sponsoring research and submitting data to EPA to be able to rely on the advice given point to a single body discharging both functions. A board that receives and reviews both pre-experiment protocols and their ultimate results would be better informed and more capable of undertaking either review.

Finally, the committee strongly recommends that the results of the board's post-experiment review be made public. The same arguments for public disclosure apply as in the case of pre-experiment review. The arguments for confidentiality are more limited, however, as the experiments have now been completed and are being voluntarily submitted to EPA for its regulatory use. It is possible that there still may be some legitimate claims of trade secrets or confidential business information that would not be publicly disclosed as an inevitable consequence of the submission to EPA. Public disclosure of the board's review should be limited as necessary to protect the sponsors' legitimate claims for protection.

Most of the procedures set forth in this chapter are prospective only, applying to experiments completed or proposals for experiments made after the recommendations of this report are implemented. Earlier experiments or later experiments not sponsored by EPA or those submitting the results to EPA also can raise scientific and ethical issues. The board should review those studies for their scientific and ethical propriety under the standards set out in Chapters 3 through 5. The board also should review those studies to offer its advice on their scientific value to the relevant EPA unit.

ONGOING ASSESSMENT AND MONITORING

The committee reached two other conclusions about the board and made one overall recommendation for review of the board itself. First, the committee believes that, over time, the board should do more than review the proposed experiments and experimental results put before it. It will have an excellent opportunity to study intentional human dosing studies and make general recommendations concerning them. On the one hand, the board might conclude, based on its experience, that such experiments, in fact, have little value or suffer from major unresolved ethical problems. On the other hand, it may be able to make specific recommendations for improving either the scientific value or the ethical propriety of these studies.

In particular, the committee strongly urges that the board should consider the issue of payments made to research participants (as discussed in Chapter 5). This is not a new issue or one unique to these studies. The ethical problems involved in paying research participants have been long recognized and debated, but they have not been resolved. The committee finds these issues particularly complicated in intentional dosing studies, in part because the altruistic motives that may inspire volunteers in other research, such as drug studies, seem more complex in the context of those involving pesticides. As the board gains experience with the payment arrangements made in various experiments and with the nature of the volunteers enrolled, it may be able to reach some valuable, empirically grounded conclusions regarding these issues.

Second, the committee urges that the board institute "ethical audits" of experiments that involve the intentional administration of toxic substances to research participants. These audits would examine a sample of these experiments to determine whether those conducting them are following, or have followed, the protocols set out in their submissions to the board. Such an audit function has been called for in general reports on protecting research participants (IOM, 2003); the strong ethical concerns and public controversy regarding human toxicant experiments make them a particularly good subject for this action.

Finally, the committee recommends a structure for review of these experiments that should be both rigorous and practical, but it recognizes its limits in foreseeing how well the structure might work over time and whether it will continue to be needed. A timely review, preferably conducted by a body including individuals from outside EPA, should help ensure that the board plays the important role this committee envisions for it.

Recommendation 6-3: Review of the Human Studies Review Board

The proposed Human Studies Review Board, its functions, and its record should be assessed after five years by a body composed of EPA staff and external reviewers.

SUMMARY

This chapter provides a procedural framework for EPA's review of intentional human dosing studies, which should be used to implement the substantive recommendations offered in previous chapters. In this chapter, the committee recommends that EPA require all human research conducted for regulatory purposes be approved in advance by an appropriately constituted IRB or an acceptable foreign equivalent. The recommendation includes EPA-conducted research; thus, EPA should ensure that research involving humans that it sponsors or conducts undergoes appropriate IRB review. EPA may want to establish its own IRB or specifically authorize another IRB to fulfill that role. Any such IRB may need special expertise to review these types of studies.

Furthermore, to assure that intentional human dosing studies conducted for EPA regulatory decisions meet the highest scientific and ethical standards, the committee recommends that EPA establish a Human Studies Review Board to address in an integrated way the scientific and ethical issues raised by intentional human dosing studies. The board should prospectively review all EPA-sponsored or EPA-conducted studies. The committee recommends that, if legally and logistically feasible, private entities that anticipate submitting the results of intentional dosing studies to EPA for regulatory purposes also should be required to submit protocols to the board before beginning a study. The submission should include the proposed protocol and sufficient background information to establish the scientific value of the experiment and provide assurance of participant safety. The proposed board supplements but does not replace review by an IRB.

The post-experiment review function of the board is distinct from the kind of review that EPA undertakes for the purpose of incorporating results from particular experiments into the regulatory process. It would not replace or modify the structures and procedures for existing EPA review. Instead, it would offer nonbinding advice to the relevant EPA units about the scientific and ethical acceptability of the submitted and completed research. The proposed review board, its functions, and its record should be assessed after five years.

REFERENCE

Institute of Medicine (IOM). 2003. *Responsible Research: A Systems Approach to Protecting Research Participants*. Washington, D.C.: The National Academies Press.

7
EPA's Use of Data from Intentional Human Dosing Studies in Risk Assessment

INTRODUCTION

As described in Chapter 1, the committee was asked by the Environmental Protection Agency (EPA) to evaluate the use of data from intentional human dosing studies in the agency's risk-assessment process. The committee examined the relevant questions within the context of EPA's general framework for risk assessment but was not asked to evaluate that framework. Rather, it was asked to determine whether and in what way data from intentional dosing studies in humans could be appropriately incorporated into EPA's established approach to risk assessment.

This chapter focuses on how EPA might use the data obtained through research that meets the scientific, ethical, and procedural standards outlined in the preceding chapters. It provides a description of the risk-assessment framework, and its basis, as the starting point for this evaluation. Following that description, the chapter examines the questions of when and how data from intentional dosing studies should be incorporated into EPA's risk-assessment process. The descriptions that follow will show that there is substantial precedent at EPA for using such data in risk assessment; however, although it is important to recognize this historical use, the committee looked beyond precedent to examine anew the questions at hand.

EPA'S RISK ASSESSMENT FRAMEWORK

Toxicity information from animal studies is routinely used by EPA and other regulatory and public health agencies to establish levels of lifetime daily intake of chemical substances that are likely to be without adverse effects in the general human population. EPA describes such levels (or doses) as risk Reference Doses (RfDs) for oral exposure and as risk Reference Concentrations (RfCs) for inhalation exposures (Faustmann and Omenn, 2001). EPA and other agencies also may establish levels to protect humans against short-term exposures or to protect specific subpopulations, such as individuals in occupational settings. In addition, EPA develops risk values for carcinogens, often based on animal data.

The use of animal data rests on substantial evidence that a relatively high degree of concordance exists between experimental animal findings and expected outcomes in humans. Their use also is necessitated by the fact that it is not possible to evaluate most forms of toxicity through intentional dosing studies in humans. The predictive value of animal studies for humans is, however, far from exact. Therefore, decisions to use results from animal studies rest, at least in part, on a "science policy" choice in which results from animal studies are generally assumed to hold for humans unless there is highly convincing evidence to the contrary (NRC, 1983; NRC, 1994).Taking into account the extensive batteries of animal studies conducted at high doses that are normally required for human risk assessment and the safety factors and conservative assumptions built into the risk-assessment process, this science policy choice generally reflects a cautious stance; but the possibility that animal studies, no matter how complete, may sometimes fail to reveal adverse health effects that are significant for humans cannot be ignored.

Data from some types of human studies have played a significant role in the establishment of RfDs, RfCs, and other measures of protection (Dourson et al., 1996; EPA, 1999). Data from both intentional dosing studies in humans and epidemiological studies have been used, with the former generally limited to effects resulting from single or very short-term exposures. In a significant number of important cases, EPA has elected to derive risk values for carcinogens from epidemiological data (e.g., benzene, arsenic, chromium [VI], and several others), given the strength of the databases for these compounds. EPA guidelines for risk assessment express a preference for human over animal data, although they clearly note the difficulties in developing human data adequate for such quantitative assessments (EPA, 2003).

Generally, knowledge of the quantitative differences between doses causing adverse effects in animals and those causing adverse effects in humans is not precise. Beginning in the 1950s, scientists in regulatory agencies began applying "safety factors" to data from animal studies to establish "acceptable daily intakes" (Lehmann and Fitzhugh, 1954). Those factors (100-fold when the animal data were derived from chronic studies) were intended to account for the possibility that humans were, on average, more sensitive to a chemical's effects than were laboratory animals, and that some humans were more sensitive than the average. This safety factor approach found broad application over the several decades following its introduction, and its use was promoted throughout the 1970s and 1980s by many committees of the National Research Council (NRC, 1983; NRC, 1994).

During the 1980s, EPA dropped the use of the term "safety factor" and substituted the concept of "uncertainty factors" (UFs). Moreover, the agency defined more completely the use of these factors in each discrete step of the process of deriving RfDs and RfCs from animal data (EPA, 2002). Several types of default UFs are used, for example, in deriving RfDs from animal data:

- UF_A: *A factor of up to 10* is used to *extrapolate* from animals to humans, to account for the possibility that humans are, on average, more susceptible to the effects of chemical exposures than are experimental animals. (The application of UF_A to animal data [a NOAEL, or no observed adverse effect level] yields a dose that should be protective for "humans of average sensitivity.")
- UF_H: *A factor of up to 10* is used to account for *variability* in response among humans, such that some members of the human population are more susceptible than the average.

If animal data are available and reflect the effects of *chronic* exposure, then the traditional lifetime RfD is derived by (1) selecting the results from the study or studies of adequate quality that show effects at the lowest dose and (2) identifying the maximum dose from that study at which no adverse effects (in relation to control animals) were observed (a NOAEL).[1] The RfD is derived as follows:

$$\text{RfD} = \frac{\text{NOAEL}}{\text{UF}_A \times \text{UF}_H}$$

[1]In recent years, EPA has occasionally implemented the concept of "Benchmark Dose" for empirically derived NOAELs. This refinement eliminates some of the variability introduced because of experimental design differences and also more clearly incorporates the notion that the experimental NOAELs are in part a function of the statistical power of experiments. None of the discussion in this chapter is influenced by these changes at EPA, and the traditional NOAEL terminology is retained for ease of presentation. It is presumed that EPA will estimate and use such Benchmark Doses in deriving RfDs (EPA, 2000b).

If EPA has available data only from subchronic studies, then another factor (UF_s) having a value up to 10 may be applied, and if the available toxicity studies do not include a NOAEL, but only an adverse effect dose (a LOAEL, or lowest observed adverse effect level), another factor (UF_L) of up to 10 may be included (Dourson et al., 1996). Thus, the derivation of an RfD from a subchronic toxicity study that reveals a LOAEL, but not a NOAEL, may proceed as follows:

$$RfD = \frac{LOAEL}{UF_L \times UF_S \times UF_A \times UF_H}$$

Such a derivation would, at a maximum, include a total UF of 10,000, although an agency technical panel has recently proposed a limit on UFs that would put a maximum total value at 3,000 (EPA, 2002). EPA sometimes includes additional factors in situations in which data deficiencies of various types exist (Faustmann and Omenn, 2001).

In connection with requirements of the Food Quality Protection Act (FQPA), EPA may include an additional safety factor to account for concerns regarding children's exposure to pesticides and susceptibility to pesticide toxicity (see Chapter 2). It should be noted that, in the case of pesticides, EPA generally does not need to include a UF_L, a UF_S, or a factor for data deficiencies, because pesticide registration requirements ensure that the agency has a complete database available. Thus, pesticide RfDs derived from animal data generally involve UF_A, UF_H, and a safety factor to protect children (see Chapter 1).

The magnitudes of the various UF_S generally range from 1 to 10. In the case of UF_A and UF_H, deviations from 10 (the typical default value) usually require substantial evidence to demonstrate that a smaller value is adequate (Dourson et al., 1996). There are many examples of regulated chemicals in which values less than 10 have been used by EPA, although 10 is typically the default in pesticide regulation. UF values less than 10 for UF_L and UF_S are common and depend on case-by-case judgment.

Historically, when human data are used as the basis for RfD derivation, the UF_A factor is replaced with actual data (e.g., a NOAEL derived from adequate human studies might be used without the need for a UF_A). The value of UF_H used in such situations depends on a judgment regarding the quality of the human data and the degree to which the studied populations represent the average or the more sensitive end of the spectrum of human sensitivities (see, for example, EPA's approach to RfD derivation in the case of methylmercury).

The scientific bases for the default values of 10 for UF_A and UF_H are limited, although several empirical studies, based on cases in which comparative data are available for animals and humans, generally have shown these values to be adequately protective. This conclusion is, however, far from fully substantiated (Rodricks et al., 2001).

IMPLICATIONS OF THE USE OF INTENTIONAL HUMAN DOSING STUDIES IN THE RISK ASSESSMENT PROCESS

Assuming that data from a given study meet the criteria for scientific validity and are found to be of adequate quality as demonstrated in Chapter 3, and assuming that they satisfy the ethical requirements described in Chapters 4, 5, and 6, they can be considered for use in risk assessment. Direct use of such human data would eliminate the need for introducing the uncertainty factor ordinarily used to extrapolate from animal data to humans (UF_A).

It must be emphasized that, even if UF_A were to be replaced with data from intentional dosing studies in humans, use of such data would have no effect on the other UFs typically used in deriving RfDs or other criteria for health protection. Specifically, the safety factor introduced under FQPA to protect children would not be affected by the replacement of UF_A with actual data.

Additional issues arise in considering the appropriateness of eliminating UF_A. Depending on the endpoints studied, the data from an intentional dosing study in humans could yield a $NOEL_{HU}$ (no observed effect level), a $NOAEL_{HU}$, a $LOEL_{HU}$ (lowest observed effect level), and perhaps a $LOAEL_{HU}$ (lowest observed adverse effect level) (see Box 7.1 for a summary of the committee's use of risk terminology for data derived from intentional human dosing studies). It is possible that an intentional human dosing study may yield a $LOEL_{HU}$ but not a $NOEL_{HU}$ or, conversely, may yield a $NOEL_{HU}$ but not a $LOEL_{HU}$. If the study yields a $LOEL_{HU}$ but not a $NOEL_{HU}$, because lower doses were not studied, a judgment would need to be made regarding the UF necessary to estimate a $NOEL_{HU}$ from the observed $LOEL_{HU}$. At least for data from intentional dosing studies in humans submitted to EPA by third parties, it would seem that there would be little basis for making such a judgment, and a repeat of the study using lower doses would be necessary to identify a $NOEL_{HU}$. This assumes that EPA will choose to conduct risk assessment using human data based on a $NOEL_{HU}$.

In those cases in which a $NOEL_{HU}$ is identified, but a $LOEL_{HU}$ is not (i.e., there is no identified effect level, adverse or not), different issues come into play. In the case of animal studies, several dose levels are used, with the expectation that a NOEL, a NOAEL, a LOAEL, and levels revealing serious toxicity will be identified (the latter levels could not be used in an intentional dosing study in humans). Ordinarily, EPA and other regulatory agencies do not use data from "NOEL-only" studies unless no other

BOX 7.1
Committee's Use of Risk Terminology for Data Derived From Intentional Human Dosing Studies

No observed effect level ($NOEL_{HU}$)

A $NOEL_{HU}$ is the highest dose or concentration at which no changes of any kind are seen relative to controls. Depending on the number of doses studied and the ability to detect the $LOEL_{HU}$, the $NOEL_{HU}$ could underestimate the actual dose that could be given without a response.

Lowest observed effect level ($LOEL_{HU}$)/ No observed adverse effect level ($NOAEL_{HU}$)

A $LOEL_{HU}$ is the lowest dose or concentration at which a biological effect that is not adverse is seen. An example of such an effect would be cholinesterase inhibition by pesticides. A small amount of cholinesterase activity has not been demonstrated to have any adverse health effects. If lower doses are not studied the $LOEL_{HU}$ could overestimate the dose that could actually elicit a response. What the committee terms a $LOEL_{HU}$ is often referred to by EPA as a no observed adverse effect level ($NOAEL_{HU}$). The committee is careful in its use of the term "$NOAEL_{HU}$" because it is most appropriately used in situations in which a clear $LOAEL_{HU}$ has been identified. A $NOAEL_{HU}$ is the highest dose or concentration at which no adverse effect is seen relative to controls.

Lowest observed adverse effect level ($LOAEL_{HU}$)

A $LOAEL_{HU}$ is the lowest dose or concentration at which an adverse effect is seen. In terms of the committee's discussion, for intentional human dosing studies there should be high confidence that any anticipated adverse effect *is not serious* and is *reversible.*

data are available and decisions must be made (typically in emergency situations; EPA, 2000a). One problem with NOEL-only studies is that they offer no information on the quantitative relationship between the measured NOEL and the unmeasured NOAEL or LOAEL, so that it is not possible to determine whether the minimum effect or toxic level is a small- or large-multiple of the measured NOEL. If the true but unidentified NOAEL or LOAEL is a large multiple of the measured NOEL, then use of the latter in deriving an RfD or similar protective value will lead to unnecessarily restrictive limits (i.e., the measured NOEL is almost certainly smaller than the "true NOEL," which should be close to the NOAEL).[2]

But a more important issue arises in connection with "NOEL-only" studies in humans ($NOEL_{HU}$-only), and that concerns the possibility that the study participants may be somehow "nonresponsive" (we are using $NOEL_{HU}$ to mean a dose producing *no response* of any type significantly different from that observed in control groups), or that there were problems with the assay employed in that study. If an intentional dosing study in humans shows no effects significantly different from the control (adverse or not), then the possibility that the volunteers chosen are somehow insensitive to the exposure or there is some other study defect that cannot be excluded, and the study should be repeated.

In general, therefore, any useful human study must investigate a range of doses, including at least one dose with an effect and one without. How many doses are studied, and how far apart they are, will determine the precision of estimates of $NOEL_{HU}$ and $LOEL_{HU}$ (or $NOAEL_{HU}$ and $LOAEL_{HU}$). In addition, the finding of an effect confirms the "assay sensitivity" of the study. A study showing no effect, and therefore providing a potentially conservative (i.e., falsely low) estimate of $NOEL_{HU}$ that might seem acceptable would be uninformative because of the lack of evidence of assay sensitivity. Thus, if an intentional dosing study in humans reveals no effects significantly different from the control (adverse or not), then the possibility that the volunteers chosen are somehow insensitive to the exposure cannot be excluded, and the study should be repeated. In general, $NOEL_{HU}$-only studies should not be used for formal risk assessments, unless no other data are available and there is a critical need to develop a tentative risk value.

ELIMINATING THE UF_A

A significant issue arising in connection with the substitution of a $NOEL_{HU}$ or $LOEL_{HU}$ for a NOAEL derived from an animal experiment concerns the matter of the uncertainty factor to be used. Obviously, the use of the traditional UF_A is not appropriate in the presence of relevant and reliable human data, because it was introduced as a default to account for the possibility that humans are, on average, more sensitive than are experimental animals. The human data replace that default assump-

[2]If the true but unidentified NOAEL is a small multiple of the NOEL, then the latter better approximates the "true NOEL"—the latter is the *maximum* dose found to produce no observed adverse effects.

tion. However, a decision to reduce or eliminate the UF_A does not automatically eliminate the uncertainty associated with using a $NOEL_{HU}$ or a $LOEL_{HU}$.

This uncertainty arises because of a certain vagueness in EPA's risk-assessment methodology—namely, the absence of a completely clear understanding of the sensitivity of the segment of the human population that is the intended target when the UF_A is applied to a NOAEL from an animal study. Thus, the use of a UF_A intended for extrapolation from animals to humans gives rise to the question: what humans? If these humans are thought to be humans of average sensitivity, then what is meant by that term? And how is it possible to know that the research participants in the intentional dosing study that is the basis for the $NOEL_{HU}$ and $LOEL_{HU}$ are truly representative of the average humans that are the intended target of the application of a UF_A? If this cannot be known, does this suggest the need for a UF_H that is somewhat larger than the usual default value of 10?

As a general matter, the risk-assessment methodology assumes that humans of average sensitivity are healthy adults, and that healthy adults are usually the participants in intentional dosing studies. But, because of the uncertainty described above, the committee considered the following possibilities:

a. Should research sponsors be encouraged to conduct two independent intentional dosing studies in humans, using different study populations and testing facilities? Replication of study results provides added confidence regarding the sensitivities of the studied populations and the degree to which those populations can be said to represent individuals of average sensitivity, while failure to replicate findings provides a measure of the variability in responses among healthy adult volunteers and a basis for assuring that risks will not be underestimated (because the data from the more sensitive study population would be used for risk assessment).

b. If research sponsors chose to conduct only a single intentional dosing study in humans, should EPA consider applying a UF_H to the $NOEL_{HU}$ that is somewhat larger than the usual factor of 10? This larger UF_H would account for the possibility that the participants in the study may be somewhat less sensitive than the hypothetical average to which the traditional UF_A is meant to apply.

The committee did not turn these two possibilities into formal recommendations because they would likely alter EPA's usual approach to risk assessment, but it concluded that they deserve study and further consideration by the agency. In some cases, volunteers are selected for intentional dosing studies specifically because they are known to be somewhat more sensitive than average. This situation occurs, for example, in some of the short-term air pollution studies conducted by or for EPA. The need for replication or additional UFs in such circumstances is significantly less compelling than those in which healthy volunteers, not known to have special sensitivities, are the participants in an intentional dosing study.

THE CASE OF CHOLINESTERASE INHIBITION

EPA's Office of Pesticide Programs has received a body of data from intentional dosing studies in humans sponsored by third parties involving measurements of cholinesterase inhibition induced by certain pesticides (see Chapter 1). Generally, such inhibition is taken to be a very sensitive marker of exposure to this class of pesticides: When RfDs are derived on the basis of NOAELs for this effect obtained from animal studies, they are generally lower than RfDs derived from studies of other adverse effects of the pesticide (including studies of chronic duration), so they are chosen as the basis for regulatory standards. (The committee heard detailed discussions and support for this position from EPA scientists and officials during its open meetings.)

There is a long history of use in pesticide regulation of $NOEL_{HU}$ or $LOEL_{HU}$ from intentional dosing studies in humans of cholinesterase inhibition, but the appropriateness of using data from such studies has come under question, and those questions gave rise to the work of this committee (see Chapter 1). The committee examined a subset of the third-party intentional dosing studies in humans submitted to EPA, and found that although these studies were not developed using the criteria for scientific validity the committee presents in this report, it appears that some of the studies may meet most of those criteria.

The full evaluation of the quality of the submitted data requires highly intense and detailed work that is beyond the scope of this committee's work and falls within EPA's regulatory responsibility. EPA is responsible for determining whether, upon close scrutiny, some or all of the submitted intentional dosing studies in humans on cholinesterase inhibition substantially meet the criteria for validity that the committee has elucidated and yield data of sufficient quality. In doing so, the agency must make clear that the particular response measured in all of these cholinesterase inhibition studies is the critical effect on which RfDs are to be based for each of the pesticides considered and that possible RfDs derived for the same pesticide—based on other findings in animals that are more relevant to chronic effects in humans—are not lower in value. If they are, they should be used rather than the RfD based on cholinesterase inhibition. Such a determination is a critical component of the criteria for scientific validity.

OTHER USES OF DATA

Data from intentional dosing studies in humans often can affect and improve risk assessments in an indirect way. For example, data from pharmacokinetic (PK) or comparative metabolism studies may be used to improve the basis for interspecies extrapolation (e.g., through development of physiologically based PK models)(Andersen, 1995). Data from such intentional dosing studies in humans are generally not used directly in risk assessments—they are not used to replace animal NOAELs, for example—but rather may improve the basis for extrapolation to humans of toxicology data obtained in animals. The increasing use of such studies is often encouraged by EPA (EPA, 2003), and their conduct does not ordinarily present the same ethical issues raised by studies in which potential adverse effects are studied (Chapters 3 and 4). Nonetheless, EPA should ensure that the scientific validity and data quality criteria described by this committee are satisfied before using this type of information in its risk-assessment process.

CONCLUSIONS AND RECOMMENDATIONS

Data from intentional dosing studies in humans that have been developed using the criteria for scientific and ethical validity elucidated in this report—and that are shown upon review to be of adequate quality—can be used within the framework for risk assessment developed by EPA. Use of such data will allow the elimination of the uncertainty factor (UF_A) ordinarily used to extrapolate from animals to humans of average sensitivity. Other uncertainty factors and the safety factor called for under FQPA to protect children are in no way affected by the use of data from the intentional human dosing studies conducted to date. It is possible that some types of metabolism and pharmacokinetics studies, together with studies of effects on the critical biomarkers, could be pertinent to the UF_H if a sufficiently broad population were studied.

To review data submitted from intentional dosing studies for regulatory decision-making purposes (e.g., setting standards), EPA should ensure the availability of sufficient and appropriate in-house expertise of at least the same level that exists for review of animal studies. The results of scientific review of data for regulatory purposes and their use in standard setting should be communicated to the Human Studies Review Board, recommended in Chapter 6. It is the committee's view that the Human Studies Review Board is advisory only and would not serve as a replacement for the scientific review EPA must perform in making regulatory decisions.

Recommendation 7-1: Review of Scientific Data

EPA's use of data from third-party intentional human dosing studies involving cholinesterase inhibition is advisable only if the agency undertakes a thorough review of the data (of the type typically undertaken for submitted animal studies and informed by external peer review) and finds that the studies substantially meet the scientific and ethical standards elucidated in this report. If the studies are found to be scientifically and ethically satisfactory, EPA should use the data to establish RfDs.

For those cholinesterase inhibitors that have been thoroughly investigated in high-quality animal studies (including studies of developmental neurotoxicity), and for which it is clear that cholinesterase inhibition is the most sensitive indicator of toxicity, data from intentional human dosing studies may be considered for use in risk assessment. It should be recognized that these circumstances—in which the most sensitive indicators of toxicity are the acute biological effects of chemicals and in which such effects are readily measurable in ethically acceptable human studies—are likely to be highly unusual. The committee's recommendations regarding the cholinesterase inhibition studies are thus not expected to suggest many other cases in which dosing studies in humans to establish a NOAEL will be of value and justifiable. The committee's recommendations regarding study justification, in which proponents of intentional dosing studies in humans must document that the endpoints to be measured are the critical determinants of risk, represent a substantial hurdle.

Recommendation 7-2: Use of Existing Cholinesterase Inhibition Studies

The cholinesterase inhibition studies that already have been submitted to EPA, if determined to be scientifically valid and justified for EPA's regulatory purposes, may be considered for use in risk assessment and standard setting if they were not unethically conducted (see Recommendation 5-7).

As discussed in Recommendation 3-1 (Chapter 3), under stringent conditions data from intentional dosing studies in humans can be used within EPA's risk-assessment framework. Use of such data may eliminate or modify the 10-fold default uncertainty factor (UF_A), ordinarily used to extrapolate from animals to humans of average sensitivity. The safety factor called for under FQPA to protect children will not be affected by the use of data from intentional dosing studies in humans that address the interspecies uncertainty factor.

Recommendation 7-3: Eliminating or Replacing the Interspecies Uncertainty Factor

In considering the use of data from the cholinesterase inhibition studies already submitted to EPA, the agency should clearly communicate to all stakeholders that information used to eliminate the interspecies uncertainty factor (UF_A) will have no influence on the use of other uncertainty factors or on the use of the safety factor protecting children as required by FQPA.

Several critical questions remain regarding the use of data from intentional dosing studies in humans. Studies that reveal "no-effects" of any type at any doses used (so-called NOEL-only studies) may provide some data regarding safety, but they are inadequate for use in deriving RfDs or any other formal measure of human protection because they provide no assurance that the study was capable of detecting the effect of interest. Such data should be used only if there are no other data available and there is a compelling public health need to derive a tentative measure of public health protection. Moreover, the relationship between the presumed sensitivity of the study population and the presumed sensitivity of average humans is somewhat ambiguous and needs clarification. Thus, it is not completely clear that the individuals that are the subjects of intentional dosing studies are always "individuals of average sensitivity" and that they are not less sensitive than the "average" individual. Uncertainties regarding these relationships may be dealt with by a requirement for study replication in a different setting or by the use of an uncertainty factor for intraspecies extrapolation (UF_H) that is somewhat greater than the usual factor of 10.

Recommendation 7-4: Data from NOEL-Only Studies and the Sensitivity of Study Populations

EPA should reject data from NOEL-only studies for risk assessments if the NOEL is defined as the absence of any biological response, because such studies do not show levels that give rise to an effect (the LOEL [lowest observed effect level]). Such studies provide no assurance that they were adequate to detect the effect of interest. The agency also should consider whether the uncertainty factor used for intraspecies variability (UF_H) should be increased to deal with the possibility that study participants may be of less than average sensitivity. A request for study replication also should be considered as a way to address this last issue.

SUMMARY

This chapter provided a description of the risk-assessment framework, and its basis, as the starting point for the committee's evaluation. The committee then examined the questions of when and how data from intentional dosing studies should be incorporated into EPA's risk-assessment process. There is substantial precedent at EPA for using such data in risk assessment. Direct use of such human data would eliminate the need for introducing the uncertainty factor ordinarily used to extrapolate from animal data to humans (UF_A). However, even if the UF_A were to be replaced with data from intentional dosing studies in human, the use of this data would have no effect on the other UFs typically used in deriving RfDs or other criteria for health protection. More specifically, the safety factor introduced under FQPA to protect children would in no way be affected by the replacement of UF_A with actual data.

To review data submitted from intentional dosing studies for regulatory decision-making purposes (e.g., setting standards), EPA should ensure the availability of sufficient and appropriate in-house expertise, at least at the level that exists for review of animal studies. The results of scientific review of data for regulatory purposes and its use in setting standards should be communicated to the Human Studies Review Board, recommended in Chapter 6.

For those cholinesterase inhibitors that have been thoroughly investigated in high-quality animal studies (including studies of developmental neurotoxicity), and for which it is clear that cholinesterase inhibition is the most sensitive indicator of toxicity, data from intentional human dosing studies may be considered for use in risk assessment. It should be recognized that these circumstances—in which the most sensitive indicators of toxicity are the acute biological effects of chemicals and in which such effects are readily measurable in human studies involving minimal risk—are likely to be highly unusual. In considering the use of data from the cholinesterase inhibition studies already submitted to EPA, the agency should clearly communicate to all stakeholders that information used to eliminate the interspecies uncertainty factor (UF_A) or to replace it with a different factor will have *no influence* on the use of other uncertainty factors or on the use of the safety factor protecting children as required by FQPA.

Several critical questions remain regarding the use of data from intentional dosing studies in humans. Uncertainties regarding these relationships may be reduced by a requirement for study replication in a different setting or by the introduction of a UF_H somewhat larger than the usual value of 10.

REFERENCES

Andersen, M. E. 1995. Development of physiologically based pharmacokinetic and pharmacodynamic models for applications in toxicology and risk assessment. *Toxicology Letter* 79:35-44.

Dourson, M. L., S. P. Feltar, and D. Robinson. 1996. Evolution of science-based uncertainty factors. *Regulatory Toxicology and Pharmacology* 24:108-120.

Environmental Protection Agency (EPA). 1999. *Policy and Procedures on Protection of Human Research Subjects in EPA Conducted or Supported Research, EPA Order.*

EPA. 2000a. *Integrated Risk Information System.* Office of Research and Development. National Center for Environmental Assessment. Available at www.epa.gov/iris.index.html.

EPA. 2000b. *Benchmark Dose Technical Guidance Document (External Review Draft).* Risk Assessment Forum. EPA/630/R-00/001.

EPA. 2002. *A Review of the Reference Dose and Reference Concentration Processes: Final Report.* EPA/630/P-02/002F.

EPA. Risk Assessment Forum. 2003. *Draft Final Guidelines for Carcinogen Risk Assessment (External Review Draft).* Risk Assessment Forum. EPA/630/P-03/001A, NCEA-F-0644A.

Faustman, E. M., and G. S. Omenn. 2001. Risk Assessment. In: Klaassen, C.D. ed. *Casarett and Doull's Toxicology: The Basic Science of Poisons.* 6th ed. New York: McGraw Hill.

Lehmann, A. J., and O. G. Fitzhugh. 1954. 100-fold margin of safety. *Quarterly Bulletin of the Association of Food and Drug Officials of the U.S.* 18:33-35.

National Research Council (NRC). 1983. *Risk Assessment in the Federal Government: Managing the Process.* Washington, D.C.: National Academy Press.

NRC. 1994. *Science and Judgment in Risk Assessment.* Washington, D.C.: National Academy Press.

Rodricks, J. V., D. W. Gaylor, and D. Turnbull. 2001. Quantitative extrapolations in toxicology. In: Hayes, A. W., ed. *Principles and Methods of Toxicology,* 365-386. London: Taylor and Francis.

Appendix A

Values and Limitations of Animal Toxicity Data

Data derived from human chemical exposure studies allow researchers to avoid many of the uncertainties and problems that are inherent in interspecies extrapolations. High-quality human data are preferred by regulatory agencies for use in assessing the potential of chemicals to cause adverse health effects in exposed populations. This is the case for the Environmental Protection Agency (EPA; 1994), the Food and Drug Administration (FDA, 2000), Health Canada (Meek et al., 1994), and the World Health Organization (IPCS, 1994). As described in Chapter 1 of this report, uncertainties associated with animal data are reflected by the routine use of a 10-fold interspecies uncertainty factor when extrapolating from laboratory animals to humans. Using existing human data for risk assessment, of course, is dependent on the quality of the data. The Food Quality Protection Act (FQPA) of 1996 specifies that there should be "reasonable certainty of no harm" occurring from pesticide residues in foods. Pertinent, scientifically valid human data should provide those assessing risk the highest degree of certainty that they are being protective but not overly conservative by relying too heavily on default approaches.

Knowledge of chemical toxicity can be gained from several types of human studies. Intentional dosing studies of humans typically involve acute or short-term administration of low to moderate doses of drugs, vaccines, cosmetics, food additives, pesticides, or occupational or environmental agents. Doses of potential therapeutic agents may be high enough to elicit adverse effects in Phase 1 clinical trials, in order to adequately characterize their tolerability. Compounds suspected or known to be toxic are commonly administered to patient volunteers rather than healthy volunteers (FDA, 2002). Doses of occupationally and environmentally encountered chemicals may also be high enough to elicit reversible biochemical, physiological, or toxicological effects.

The intentional human dosing studies with pesticides reviewed by this committee involved low doses that produce no effects or minimal, reversible changes in sensitive biomarkers, albeit in one study the effect was sufficiently large to warrant termination of the study. Although epidemiological investigations of exposed populations may identify associations of adverse effects and chemical exposures and support inferences of cause and effect, epidemiological data are nonetheless usually limited by inadequate characterization of exposures and by an inability to recognize or control confounding factors (Dourson et al., 2001). Most clinical case reports of toxicant exposures have the same limitations. Such information, however, can alert us to previously unrecognized toxicities and identify critical effects to evaluate in subsequent investigations. Human cells and tissues can be very useful for metabolism and mode of action studies (MacGregor et al., 2001). Good correlation is often found between the metabolism of chemicals in vivo and metabolism by isolated hepatocytes of the same species (Oesch and Diener, 1995). Mechanistic studies with humans and laboratory animals may identify relevant toxicity end points and bioactive moieties and facilitate development of the most pertinent animal models (Jorkasky, 1998; Gregus and Klaassen, 2001).

Toxicological data from human exposure to pesticides and other chemicals are often limited or nonexistent. Obviously, one cannot administer sufficient amounts of a chemical to characterize the dose dependency of major adverse effects that exposed individuals could experience. Long-term exposures cannot be conducted in order to elicit chronic conditions. Parallel laboratory animal-human experiments, however, can be very useful in assessing the relevance of particular animal models to humans. Ideally, toxicologists and risk assessors would like to have dose-response data from experiments in which the same parameters were monitored and in which there was overlap of the range of doses given to each species. The doses administered to humans would be relatively low, but they should produce changes in sensitive adaptive effects, or biomarkers. Pharmacokinetic (PK), metabolic, and mechanistic studies in humans and animals also provide valuable information for scientifically based interspecies extrapolations (Jorkasky, 1998). Nonetheless, comprehensive toxicology investigations in different species of laboratory animals are necessary to fully evaluate the hazard potential of most chemicals.

Evaluation of the toxicity of chemicals in laboratory animals is a cornerstone of human safety evaluation. Experimentation with animals makes it possible to learn a great deal about the toxic potential of drugs and other chemicals. Explicitly defined investigations in laboratory animals are prescribed by EPA, FDA, and other regulatory authorities for approval of pesticides and drugs. Animals can be utilized for short, intermediate, and chronic exposure studies through which scientists can characterize the spectrum of adverse effects of a compound over a wide range of doses, dosage regimens, and exposure durations. Often, the toxicologist initially will administer high doses and evaluate a broad spectrum of parameters in order to identify target organs. Focused dose-response studies employing a limited number of sensitive indices of injury can then be performed. Ideally, dosage routes and regimens will be designed to mimic actual human exposure situations. The use of laboratory animals as toxicology research subjects is advantageous for several reasons. Most rodent species are relatively inexpensive and easily maintained. Large numbers of rodents can be assessed over a wide range of doses, increasing the likelihood of detecting adverse events (Zbinden, 1991). A number of biochemical, cellular, and physiological endpoints that can be examined only in human biopsy samples or at autopsy can be evaluated in animals. In addition, considerable background information often is available on commonly used strains of mice, rats, and dogs, including their genetic makeup, their abilities to metabolize xenobiotics, and their responses to other compounds. Groups of uniform animals can be administered measured doses of chemical(s) under defined and carefully controlled conditions, circumstances under which adverse effects to a specific chemical exposure can be attributed with greater certainty. Human populations, in contrast, are much more genetically diverse (Weber, 1999), with endogenous and exogenous factors (e.g., diet, stress, health, age, personal habits, use of drugs, exposures to other chemicals) that may not be recognized or controllable. In addition, the degree and duration of an individual's exposure to the chemical of interest are often unclear in epidemiological studies and case histories.

Findings in animal toxicology studies generally are applicable to humans, although responses of laboratory animals and humans to chemicals may differ qualitatively and/or quantitatively. The most definitive study to date of interspecies concordance involved an International Life Sciences Institute-sponsored review of data supplied by 12 pharmaceutical companies (Olson et al., 2000). The database consisted of toxicity findings from preclinical (i.e., experimental animal) and clinical (i.e., human) studies of 150 compounds in 15 therapeutic classes. Interspecies concordance of toxicity was said to exist if generally severe effects on the same organ occurred in humans and in laboratory animals. There was an overall interspecies concordance for 61 percent of the compounds. Rodents alone were predictive of human toxicities for 43 percent of the agents, while nonrodents (primarily dogs) alone were predictive for 63 percent. In another comparative investigation, 43 percent of the clinical toxicities of 64 marketed drugs in Japan were not predicted from animal experiments (Igarashi, 1994). The poorest concordance in this and the Olson survey (2000) were for cutaneous hypersensitivity and endocrine and hepatic functions. Obviously, animal studies cannot reveal subjective effects such as headache, myalgias, dizziness, nausea, or mental disturbances. The Olson study described other reports of poor correspondence between animal data and human toxicities severe enough to lead to market withdrawal of drugs. Many of these cases apparently involved idiosyncratic reactions that occurred with a very low incidence in patient populations, a phenomenon that reflects the pronounced influence of exogenous and endogenous factors on interindividual responses.

An evaluation by Dourson et al. (2001) of susceptibilities to industrial and agricultural chemicals has provided some additional information on the reliability of animal toxicology findings. These investigators compared human data-based reference doses (RfDs) for 22 chemicals in EPA's Integrated Risk Information System (IRIS) database with RfDs the authors calculated from animal data in IRIS using standard uncertainty factors. Seven of the 22 compounds were pesticides, for which cholinesterase inhibition was measured in intentionally dosed research participants. The interspecies concordance rate was approximately 40 percent. The human-based RfDs were lower than the animal-based values for 7 (32 percent) of the 22 chemicals. The human values were more than three times lower for five of these seven compounds, leading the authors to conclude that exposure limits based upon animal data may not be protective of public health. The power of Dourson's analysis is somewhat limited by the modest number of chemicals that were evaluated and by the quality and applicability of some of the data.

A considerable amount of information has been published on interspecies similarities and differences in susceptibility of chemical carcinogenesis. Faustman and Omenn (2001) pointed out that all human carcinogens that have been adequately tested in animals have produced cancer in at least one animal model. However, an evaluation of National Toxicology Program cancer bioassay data for 400 chemicals revealed that only 23 percent of the carcinogenic compounds produced tumors in both mice and rats (Fung et al., 1995). Some carcinogens, such as vinyl chloride, produce tumors in humans and in both sexes of other species tested. Conversely, many other carcinogens appear to be sex, strain and/or species specific (Grisham, 1996). Unleaded gasoline-induced kidney toxicity and cancer, for example, are limited to male rats, which is attributed to binding of gasoline to $\propto_{2u}$-globulin, a male rat-specific protein. The protein is hypothesized to accumulate to toxic levels in kidney cells and thereby induce sustained cellular proliferation, with its attendant cancer risk factors (Lehman-McKeeman, 1997). It also is hypothesized that oxidative moieties produced by peroxisomal enzymes and modification of cell signaling by activation of peroxisome proliferator-activated receptor-α can elicit liver cancer (Lake, 1995). A variety of compounds, including drugs such as ciprofibrate and nafenopin and solvents such as trichloroethylene and perchloroethylene, markedly induce hepatic peroxisomes and produce hepatic tumors in mice and/or rats. Studies of humans taking clofibrate and gemfibrozil, however, reveal little peroxisome proliferation and no increased incidence of liver cancer. Pharmacodynamic differences (i.e., disparities in receptor numbers and affinities) appear to account for this phenomenon (Cattley et al., 1998).

Variances in pharmacokinetics are often responsible for pronounced interspecies differences in susceptibility to toxic agents. The term "pharmacokinetics" encompasses systemic absorption, distribution, metabolism, and elimination. Many chemicals undergo metabolic activation (i.e., are metabolized to toxic metabolites). Others are detoxified through metabolism. Aflatoxin B_1, one of the most potent hepatocarcinogens known, is metabolically activated by cytochrome P450s and subsequently detoxified by conjugation with glutathione. Mice have been found to be much more resistant to aflatoxin B_1-induced liver cancer than rats. This disparity has been attributed to very efficient conjugation of the major reactive metabolite by mice.

Interspecies extrapolations on the basis of body surface area and comparative metabolism studies with primary hepatocytes of mice, rats, and humans indicate that the susceptibility of humans to a number of compounds resembles that of rats (Hengstler et al., 1999). Tamoxifen is a nonsteroidal antiestrogen that is used to treat pre- and postmenopausal women with breast cancer. It is a full estrogen in mice, a partial estrogen/antiestrogen in rats and humans, and an antiestrogen in chicks (Jordan and Robinson, 1987). Tamoxifen is metabolically activated to a DNA-binding metabolite by a combination of Phase I and II metabolism. Biotransformation of tamoxifen is qualitatively similar in rats and humans, but the amounts of reactive metabolites and DNA adducts formed in the human liver are much lower than those formed in rats (Hengstler et al., 1999). Knowledge of qualitative and quantitative species differences in the metabolism of a xenobiotic allows the selection of the animal strain and species that is most like the human.

There are a number of quantitative methods for extrapolation of animal toxicity data to humans. The standard uncertainty factor default approach (described in Chapter 6 of this report) is frequently used because of a paucity of data. Linear extrapolations based on body weight equivalence often are inaccurate unless species-specific conversion factors are applied (Voisin et al., 1990), while allometric scaling on the basis of body surface area is more accurate. Freireich et al. (1966) report that doses of anticancer drugs lethal to 10 percent of rodents and maximally tolerated doses (MTDs) in nonrodents correlate with MTDs in human patients, when the doses are normalized to the surface area of each species. Normalization of body weight to the 2/3 or 3/4 power results in accurate predictions of body surface area, since both size (weight) and form (height) are taken into account (Davidsohn et al., 1986). FDA (2002) describes the use of standard species-specific factors that allow conversion of animal doses in mg/kg to animal doses in mg/m^3 and human doses in mg/kg. The use of PK and metabolism data, when available for each species of interest, facilitates the most reliable interspecies conversions.

FDA (2002) has published a draft guidance document that describes a strategy recommended for deriving safe starting doses of therapeutic agents for clinical trials with healthy research participants. The first step in the process involves the identification of NOAELs (no observed adverse effect levels) from animal toxicity studies. The NOAEL for the most appropriate species is selected, regardless of whether this species is the most sensitive. The selection is based on information available on relative bioavailability, metabolic profile, molecular biology, physiology, and reactions to similar compounds. Humans and marmosets, for example, have constitutive levels of hepatic CYP1A2, a P450 isozyme that activates heterocyclic amines to reactive metabolites (Hengstler et al., 1999). Cynomolgus monkeys lack constitutive CYP1A2. Marmosets are thus a more suitable animal model for heterocyclic amines than cynomolgus monkeys. For drugs, the most appropriate animal NOAEL is converted to the human equivalent dose (HED) by the body surface area normalization process described by FDA (2002). Finally, the HED is divided by a safety factor to yield the maximum recommended starting dose.

Pharmacokinetics-based conversions provide the most reliable means of extrapolating from one species to another. Such approaches require PK data for each species of interest. Optimally, animal blood and target organ time-course data and metabolic information will be available for a range of doses, including those within which toxicity occurs. Human metabolic and blood-level data for low doses also would be necessary. Blood time-course data alone allow comparison of areas under blood concentration versus time curves (AUCs) for test animals and humans. Physiologically-based PK (PBPK) models (described below) are more precise, versatile, and scientifically credible than classical compartment-based models for inter-route, interdose and interspecies extrapolations (Voisin et al., 1990). PBPK models incorporate the unique anatomical, physiological, and metabolic characteristics of each species, as well as the physicochemical properties of the toxicant. PBPK models can be utilized to predict blood and target organ peak concentrations and AUCs of toxic moieties, whether they are the parent compound or a particular metabolite (Gerlowski and

Jain, 1983). Toxicant exposures required to produce target organ doses that result in toxic effects of a given magnitude in laboratory animals are determined experimentally and modeled. The PBPK model is then allometrically scaled up to humans, or human-specific physiological and biochemical parameters are utilized for the model. Low-dose PK studies in volunteers are necessary to validate (i.e., assess the accuracy of) the model's predictions. Metabolic rate constants can be obtained from these studies or from in vitro experiments with human tissues or cells. Validated models allow one to simulate the human exposure conditions that will produce a target organ dose equivalent to that previously found to be associated with toxicity in the test animal. This so-called HED approach has been used successfully for a number of chemicals including, among others, methylene chloride (Andersen et al., 1991), acrylic acid (Frederick et al., 1998), and chlorpyrifos (Timchalk et al., 2002). Sensitivity analyses can be conducted to learn which physiological or biochemical parameters have the greatest impact on the pharmacokinetics of a particular chemical. One can also determine the influence of variability (that may exist in a human population) of the key parameters on estimates of tissue doses. Monte Carlo sampling of parameter distributions generates a distribution of model-generated target organ doses for different exposure regimens. The risk assessor can assess the variability in this distribution and judge whether a 10-fold intraspecies factor is merited (Watanabe et al., 1992; Thomas et al., 1996).

REFERENCES

Andersen, M. E., H. J. Clewell III, M. L. Gargas, M. G. McNaughton, R .H. Reitz, R. Nolan, and M. McKenna. 1991. Physiologically based pharmacokinetic modeling with dichloromethane, its metabolite carbon monoxide, and blood carboxyhemoglobin in rats and humans. *Toxicology and Applied Pharmacology* 108:14-27.

Cattley, R. C., J. DeLuca, C. Elcombe, P. Fenner-Crisp, B. G. Lake, D. S. Marsman, T. A. Pastoor, J.A. Popp, D.E. Robinson, B. Schwetz, J. Tugwood, and W. Wahli. 1998. Do peroxisome proliferating compounds pose a hepatocarcinogenic hazard to humans? *Regulatory Toxicology and Pharmacology* 27:47-60.

Davidsohn, I. W. F., J. C. Parker, and R. B. Beliles. 1986. Biological basis for extrapolation across mammalian species. *Regulatory Toxicology and Pharmacology* 6:211-237.

Dourson, M., M. E. Anderson, L. S. Erdreich, and J. A. MacGregor. 2001. Using human data to protect the public's health. *Regulatory Toxicology and Pharmacology* 33:234-256.

Environmental Protection Agency (EPA). 1994. *Methods for Derivation of Inhalation Reference Concentrations and Application of Inhalation Dosimetry*. EPA/600/8-90-066F. Washington, D.C.: Office of Health and Environmental Assessment.

Faustman, E. M., and G. S. Omenn. 2001. Risk assessment. In: C. D. Klaassen, ed. *Casarett and Doull's Toxicology: The Basic Science of Poisons*. 6th ed. New York: McGraw-Hill.

Food and Drug Administration (FDA). 2002. *Guidance for Industry and Reviewers. Estimating the Safe Starting Dose in Clinical Trials for Therapeutics in Adult Healthy Volunteers*. Draft. Rockville, MD: FDA.

Frederick, C., M. L. Bush, L. G. Lomax, K. A. Black, L. Finch, J. S. Kimbell, K. T. Morgan, J. B. Morris, R. P. Subramaniam, and J. S. Ultman. 1998. Application of a hybrid computational fluid dynamics and physiologically based inhalation model for interspecies dosimetry extrapolation of acidic vapors in the upper airways. *Toxicology and Applied Pharmacology* 152:211-231.

Freireich, E. J., E. A. Gehan, D. P. Rall, L. H. Schmidt, and H. E. Skipper. 1966. Quantitative comparison of toxicity of anticancer agents in mouse, rat, hamster, dog, monkey, and man. *Cancer Chemotherapy Reports* 50:219-244.

Fung, V. A., J. C. Barrett, and J. Huff. 1995. The Carcinogenesis Bioassay in Perspective: Application in Identifying Human Cancer Hazards. *Environmental Health Perspective* 103: 680-683.

Gerlowski, L. E., and R. K. Jain. 1983. Physiologically based pharmacokinetic modeling: principles and applications. *Journal of Pharmaceutical Sciences* 72:1103-1127.

Gregus, Z., and C. D. Klaassen. 2001. Mechanisms of toxicity. In: C. D. Klaassen, ed. *Casarett and Doull's Toxicology: The Basic Science of Poisons*. 6th ed. New York: McGraw-Hill.

Grisham, J. W. (1996). Interspecies comparison of liver carcinogenesis: implications for cancer risk assessment. *Carcinogenesis* 18:59-81.

Hengstler, J. G., B. van der Burg, P. Steinberg, and F. Oesch. 1999. Interspecies differences in cancer susceptibility and toxicity. *Drug Metabolism Reviews* 31:917-970.

Igarashi, T. 1994. The duration of toxicity studies required to support repeated dosing in clinical investigation: a toxicologist's opinion. In: Parkinson, C., M. McAuslane, C. Lumley, and S. R. Walker, eds. *CMR Workshop: The Timing of Toxicological Studies to Support Clinical Trials*, 55-67. Lancaster, United Kingdom: Quay.

International Programme on Chemical Safety (IPCS). 1994. Derivation of guidance values for health-based exposure limits. In: *Environmental Health Criteria No. 170: Assessing Human Health Risks of Chemicals*. Geneva: World Health Organization.

Jordan, V. C., and S. P. Robinson. 1987. Species-specific pharmacology of antiestrogens: role of metabolism. *Federation Proceedings* 46:1870-1874.

Jorkasky, D. K. 1998. What does this clinician want to know from the toxicologist? *Toxicology Letters* 102-103:539-543.

Lake, B. G. 1995. Peroxisome proliferation: current mechanisms relating to non-genotoxic carcinogenesis. *Toxicology Letters* 82-83:673-681.

Lehman-McKeeman, L. D. 1997. α_{2u}-Globulin nephropathy. In: Sipes, I. G., C. A. McQueen, and A. J. Gandolfi, eds. *Comprehensive Toxicology*, Vol. 7, 677-692. United Kingdom: Elsevier, Oxford.

MacGregor, J. T., J. M. Collins, Y. Sugiyama, C. A. Tyson, J. Dean, L. Smith, M. Andersen, R. D. Curren, J. B. Houston, F. F. Kadlubar, G. L. Kedderis, K. Krishnan, A. P. Li, R. E. Parchment, K. Thummel, J. E. Tomaszewski, R. Ulrich, A. E. Vickers, and S. A. Wrighton. 2001. *In vitro* human tissue models in risk assessment: report of a consensus building workshop. *Toxicological Sciences* 59:17-36.

Meek, M. E., R. Newhook, R. G. Liteplo, and V. C. Armstrong. 1994. Approach to assessment of risk to human health for priority substances under the Canadian Environmental Protection Act. *Environmental Carcinogenesis and Ecotoxicology Reviews* C12:105-134.

Oesch, F., and B. Diener. 1995. Cell systems for use in studies on the relationship between foreign compound metabolism and toxicity. *Pharmacology and Toxicology* 76:325-327.

Olson, H., G. Betton, D. Robinson, K. Thomas, A. Monro, L. P. Kolaja, J. Sanders, G. Sipes, W. Bracken, M. Dorato, K. Van Deun, P. Smith, B. Burger, and A. Heller. 2000. Concordance of the toxicity of pharmaceuticals in humans and in animals. *Regulatory Toxicology and Pharmacology* 32:56-67.

Thomas, R. S., P. L. Bigelow, T. J. Keefe, and R. S. H. Yang. 1996. Variability in biological exposure indices using physiologically-based pharmacokinetic modeling and Monte Carlo simulation. *American Industrial Hygiene Association Journal* 57:23-32.

Timchalk, C., R. J. Nolan, A. L. Mendrala, D. A. Dittenber, K. A. Brzak, and J. L. Mattsson. 2002. A physiologically based pharmacokinetic and pharmacodynamic (PBPK/PD) model for the organophosphate insecticide chlorpyrifos in rats and humans. *Toxicological Sciences* 66:34-53.

Voisin, E. M., M. Ruthsatz, J. M. Collins, and J. C. Hoyle. 1990. Extrapolation of animal toxicity to humans: interspecies comparisons in drug development. *Regulatory Toxicology and Pharmacology* 12:107-116.

Watanabe, K., F. Y. Bois, and L. Zeise. 1992. Interspecies extrapolation: a reexamination of acute toxicity data. *Risk Analysis* 12:301-310.

Weber, W. W. 1999. Populations and genetic polymorphisms. *Molecular Diagnosis* 4:299-307.

Zbinden, G. 1991. Predictive value of animal studies in toxicology. *Regulatory Toxicology and Pharmacology* 14:167-177.

Appendix B

Pharmacokinetics and Metabolism of Pesticides

Pharmacokinetic data are important in considering the relative risks posed by pesticides to the health of different species of laboratory animals and humans. A basic tenet of toxicology is that toxic effects are a function of the concentration of the bioactive form of a chemical in a target organ. Thus, the degree and duration of a toxic response are dependent on how much of the bioactive moiety reaches its target site and how long it remains there. This is a function of the extent of the chemical's system absorption, distribution, metabolism, interaction with cellular components, and elimination.

Dermal exposure, inhalation, and ingestion are the primary routes of human exposure to pesticides and other chemicals. The percutaneous absorption of pesticides varies widely, as members of many different chemical classes are used to control unwanted insects, fungi, plants, and animals. The outermost layer of the epidermis, the stratum corneum, serves as the barrier to penetration. The thickness of this layer over different parts of the body varies significantly, as does the extent of systemic absorption. The stratum corneum is composed largely of tightly adhering, cornified epithelial cells impregnated with sebum and sweat. The lipophilic sebum normally predominates, so as a general rule, lipid-soluble compounds are absorbed more readily than hydrophilic compounds. Shah et al. (1987), however, did not find good correlation between octanol-water partition coefficients and dermal absorption of a diverse group of 14 pesticides in rats. Solubilization of pesticides on the skin's surface greatly aids in their dermal penetration. The EPA usually requires percutaneous absorption studies in rodents as part of the pesticide registration application process. There are pronounced interspecies differences in the thickness of the stratum corneum, dermal blood flow rate, and other determinants of absorption (Mattie et al., 1994; Monteiro-Riviere et al., 1990). Human skin usually is less permeable than rodent skin to many chemicals (Poet, 2000).

Information on the absorption of pesticides and other chemicals from the lungs is often quite limited. Pulmonary exposure is not a major concern for many compounds that have low vapor pressures. Some pesticides, such as soil and grain fumigants, however, are relatively volatile and may be inhaled in high concentrations. Inhaled fumigants such as ethylene dibromide, trichloropropane, and dibromochloropropane are well absorbed and can exert toxic and/or carcinogenic effects in mice and rats. These lipophilic compounds readily diffuse across the respiratory epithelium of the alveoli into the profuse capillary supply of the pulmonary circulation. Systemic absorption of volatile organic chemicals (VOCs) (e.g., trichloroethylene) is often greater in rodents than in humans subjected to equivalent inhalation exposures (Fisher, 2000). The interspecies difference can be attributed to rodents' higher respiratory rates, cardiac outputs, and blood-air partition coefficients—three major determinants of pulmonary absorption of VOCs (Bruckner and Warren, 2001).

The gastrointestinal (GI) tract is the major portal of entry of most pesticide contaminants of food and water. The rapidity and extent of systemic absorption depends on the physical and chemical properties of the compound, as well as conditions within the GI tract. Some of the more important endogenous factors include gastric emptying and intestinal motility; gut flora; acid and enzyme secretory activities; cellular transport systems; blood supply; and mucosal structure and surface area. The small intestine has the greatest surface area and is frequently the optimal absorption site. Systemic absorption of different classes of pesticides varies widely. As a rule, lipid soluble, unionized forms are relatively well absorbed throughout the GI tract. The molecular weight of highly lipophilic compounds such as pyrethroid insecticides (Anadon et al., 1996) can increase to a point that mucosal penetration diminishes. Ingested arsenic, copper, cadmium, and other metals are poorly absorbed by adults. Experiments by Kostial et al. (1978) reveal substantially lower blood levels of lead, mercury, cadmium, and manganese in adult rats than in sucklings given comparable oral doses of the metals. Morphological and functional immaturities of intestinal epithelial cells account for the greater penetrability of the gut of neonatal animals and humans.

Once a chemical has entered the bloodstream, it is distributed throughout the body. The initial phase of systemic distribution is governed largely by tissues' rate of blood perfusion and by the rate at which the compound exits the bloodstream (Rozman and Klaassen, 2001). Certain organs, including the brain and testes, are afforded some degree of protection from polar and/or large molecules by tight capillary cell junctions, enveloping cells and transporters. The immature blood-brain barrier of young animals and children is more permeable than that of adults to metals (e.g., mercury, lead, cadmium) (Kostial et al., 1978). Some pesticides, such as dieldrin and atrazine (McMullin et al., 2003), bind extensively to plasma proteins. As long as the compounds are bound, they are not able to leave the bloodstream and reach sites of action or elimination. Substantial binding thereby generally reduces the maximum activity of chemicals, but can prolong their effects. The final phase of distribution is governed largely by the affinity of a compound or its metabolite(s) for a particular organ or tissue. The liver and kidney have a high capacity for binding a wide variety of xenobiotics. The lungs preferentially bind and accumulate paraquat, which exerts its injurious effects there. Metallothionein, a protein that avidly binds heavy metals, is present in high concentrations in the kidneys. Lipophilic pesticides, such as chlordane, DDT and pyrethroids, accumulate in body fat, from which they are slowly released (Jandacek and Tso, 2001).

Biotransformation plays a major role in preventing the accumulation of lipid-soluble xenobiotics in the body. Elimination of such compounds often depends on their conversion by enzyme-catalyzed reactions to more water-soluble forms that can be excreted in the urine and bile. Xenobiotic-metabolizing enzymes tend to have broad, overlapping substrate specificities. Many such enzymes are expressed constitutively (i.e., are synthesized in the absence of an apparent external stimulus), with the synthesis of some induced (i.e., stimulated) by the presence of the chemical they transform. Enzymes frequently exist in multiple forms (i.e., isozymes) with different substrate affinities. Xenobiotic-metabolizing enzymes and their isozymes are distributed widely throughout the body. The preponderance are found in the liver, though certain cell types in different extrahepatic organs exhibit relatively high levels of specific enzymes. There are often considerable interspecies differences in the presence and activity of enzymes and isozymes in particular tissues (Lin, 1995).

Biotransformation is a key determinant of the toxicity of many pesticides and other chemicals. Biotransformation results in detoxification and hastened elimination of some pesticides. The parent compound, for example, is responsible for the neurotoxic action of pyrethroids. These compounds are inactivated by the concerted actions of carboxylesterases and P450s-catalyzed hydroxylation and subsequent conjugation (Soderlund et al., 2002). Organophosphates are also detoxified by esterase-catalyzed hydrolysis, although desulfuration by P450s produces oxons, the neurotoxic moieties that bind to and inhibit acetylcholinesterase (Sultatos, 1994). The pronounced acute toxicity of chlorpyrifos in immature rats is attributed to their deficiencies of chloropyrifos-oxonase (i.e., the A-esterase that hydrolyzes the oxon) (Mortensen et al., 1996) and of carboxylesterase (Moser et al., 1998). Thus, recognition of metabolic differences is essential to understanding variances in the toxicity of xenobiotics in different cells, tissues, species, strains, sexes, races, and age groups.

Toxicants and their metabolites are eliminated from the body by several routes. Many xenobiotics, as described above, are converted to more water-soluble products, so that they may be discharged in the largely aqueous urine and bile. Renal excretion is the principal pathway for elimination of these compounds (Kozman and Klaassen, 2001). Biliary excretion also can play a major role for some parent compounds and metabolites, notably conjugates formed in Phase II reactions. The relative contribution of urinary and biliary excretion, and the extent of enterohepatic recirculation, are compound and species specific (Lin et al., 1995). Volatile parent compounds and metabolites can be exhaled, but this route of elimination is not important for most pesticides. Hair, fingernails, desquamated skin, and body secretions (e.g., milk, tears, saliva, and sweat) have limited capacity to eliminate chemicals.

REFERENCES

Anadon, A., M. R. Martinez-Larranaga, M. L. Fernandez-Cruz, M. J. Diaz, M. C. Fernandez, and M. A. Martinez. 1996. Toxicokinetics of deltamethrin and its 4'-HO-metabolite in the rat. *Toxicology and Applied Pharmacology* 141:8-16.

Bruckner, J., and D. A. Warren. 2001. Toxic effects of solvents and vapors. In: Klaassen, C., ed. *Casarett and Doull's Toxicology: The Basic Science of Poisons*. 6th ed. New York: McGraw-Hill.

Fisher, J. W. 2000. Physiologically based pharmacokinetic models of trichloroethylene and its oxidative metabolites. *Environmental Health Perspective* 108(Suppl. 2):265-273.

Jandacek, R. J., and P. Tso. 2001. Factors affecting the storage and excretion of toxic lipophilic xenobiotics. *Lipids* 36:1289-1305.

Kostial, K., D. Kello, S. Jugo, I. Rabar, and T. Maljkovic. 1978. Influence of age on metal metabolism and toxicity. *Environmental Health Perspectives* 25:81-86.

Lin, J. H. 1995. Species similarities and differences in pharmacokinetics. *Drug Metabolism and Disposition* 23:1008-1020.

Mattie, D. R., J. H. Grabau, and J. N. McDougal. 1994. Significance of the dermal route of exposure to risk assessment. *Risk Analysis* 14:277-284.

McMullin, T. S., J. M. Brzezicki, B. K. Cranmer, J. D. Tessari, and M. E. Andersen. 2003. Pharmacokinetic modeling of disposition and time course studies with carbon 14 atrazine. *Journal of Toxicology and Environmental Health* 66:941-964.

Monteiro-Riviere, N. A., D. G. Bristol, T. O. Manning, R. A. Rogers, and J. E. Riviere. 1990. Interspecies and interregional analysis of the comparative histological thickness and laser doppler blood flow measurements at five cutaneous sites in nine species. *Journal of Investigative Dermatology* 95:582-586.

Mortensen, S. R., S. M. Chanda, M. J. Hooper, and S. Padilla. 1996. Maturational differences in chlorpyrifos-oxonase activity may contribute to age related sensitivity to chlorpyrifos. *Toxicology Science* 11:279-287.

Moser, V. C., S. M. Chanda, S. R. Mortensen, and S. Padilla. 1998. Age and gender related differences in sensitivity to chlorpyrifos in the rat reflect developmental profiles of esterase activities. *Toxicology Science* 46:211-222.

Poet, J. S. 2000. Assessing dermal absorption. *Toxicology Science* 58:1-2.

Rozman, K. K., and C. D. Klaassen. 2001. Absorption, distribution, and excretion of toxicants. In: Klaassen, C., ed. *Casarett and Doull's Toxicology: The Basic Science of Poison*. 6th ed. New York: McGraw-Hill.

Shah, P. V., H. L. Fisher, M. R. Sumler, R. J. Monroe, N. Chernoff, and L. L. Hall. 1987. Comparison of the penetration of pesticides through the skin of young and adult rats. *Journal of Toxicology and Environmental Health* 21:353-366.

Soderlund, D. M., J. M. Clark, L. P. Sheets, L. S. Mullin, V. J. Piccirillo, D. Sargent, J. T. Stevens, and M. L. Weiner. 2002. Mechanisms of pyrethriod neurotoxicity: implications for cumulative risk assessment. *Toxicology* 171:3-59.

Sultatos, L. G. 1994. Mammalian toxicology of organophosphorus pesticides. *Journal of Toxicology and Environmental Health* 43:271-289.

Appendix C

Biographical Sketches of the Members and Staff of the Committee on the Use of Third Party Toxicity Research with Human Research Participants

Cochair, **James F. Childress (IOM),** B.A., Guilford College; B.D., Yale Divinity School; M.A. and Ph.D., Yale University, is the John Allen Hollingsworth Professor of Ethics and Professor of Medical Education at the University of Virginia, where he teaches in the Department of Religious Studies and is Director of the Institute for Practical Ethics and Public Life. He served as Chair of the Department of Religious Studies, 1972-1975 and 1986-1994, as Principal of the University of Virginia's Monroe Hill College from 1988 to 1991, and as co-director of the Virginia Health Policy Center 1991-1999. In 1990 he was named Professor of the Year in the state of Virginia by the Council for the Advancement and Support of Education. He is the author of numerous articles and several books in biomedical ethics, including *Principles of Biomedical Ethics* (with Tom L. Beauchamp); *Priorities in Biomedical Ethics; Who Should Decide? Paternalism in Health Care;* and *Practical Reasoning in Bioethics*, along with articles and books in other areas of ethics. Childress was Vice Chair of the national Task Force on Organ Transplantation, and he has also served on the Board of Directors of the United Network for Organ Sharing (UNOS), the UNOS Ethics Committee, the Recombinant DNA Advisory Committee, the Human Gene Therapy Subcommittee, the Biomedical Ethics Advisory Committee, and several Data and Safety Monitoring Boards for National Institutes of Health Clinical Trials. He was a member of the presidential-appointed National Bioethics Advisory Commission 1996-2001. Childress is a fellow of the American Academy of Arts and Sciences and, in 1998, was elected to membership in the Institute of Medicine of the National Academy of Sciences. He is also a fellow of the Hastings Center. He has been the Joseph P. Kennedy, Sr., Professor of Christian Ethics at the Kennedy Institute of Ethics at Georgetown University (1975-1979) and a Visiting Professor at the University of Chicago Divinity School and Princeton University.

Cochair, **Michael R. Taylor,** B.A. (Political Science), Davidson College; J.D., University of Virginia, is Senior Fellow and Director, Risk, Resource, and Environmental Management Division, Resources for the Future (RFF); and a member of the Board of Trustees of Resolve, Inc., a nonprofit environmental and public health mediation and dispute resolution organization. At RFF, Taylor leads a research program on the policy and institutional issues affecting the success of the global food and agricultural system in areas such as food security in developing countries, food safety as a global concern, and the natural resource and environmental sustainability of agriculture. Publications include *Redesigning Food Safety: Using Risk Analysis to Build a Better Food Safety System* (2001) (co-author). Prior to coming to RFF, Taylor served in government, practiced law in Washington, D.C., and worked in private industry. He was Administrator of the Department of Agriculture's Food Safety and Inspection Service; Deputy Commissioner for Policy at the Food and Drug Administration (FDA), and an FDA staff lawyer and Executive Assistant to the FDA Commissioner. He practiced food and drug law and was a partner in the law firm of King and Spalding and was Vice President for Public Policy at Monsanto Company. He is currently a member of the National Academies Committee on Implications of Dioxin in the Food Supply, and he has served on the Subcommittee on Defining Science-Based Concerns Associated with Products of Animal Biotechnology; the Food Forum; and the Committee on Scientific and Regulatory Issues Underlying Pesticide Use Patterns and Agricultural Innovation.

James V. Bruckner, B.S. (Pharmacy), University of Texas, Austin; M.S. (Toxicology), University of Texas at Austin; Ph.D. (Toxicology), University of Michigan, Ann Arbor, is Professor of Pharmacology and Toxicology, Department of Pharmaceutical and Biomedical Sciences, College of Pharmacy, University of Georgia. He was director of the University of Georgia's Interdisciplinary Graduate Program in Toxicology for 15 years. He was recently a member of the Federal Insecticide, Fungicide, and Rodenticide Act Scientific Advisory Panel for Evaluation of Exposure and Hazards to Children from Contact with Chromated Copper Arsenate-Treated Wood Structures, Office of Pesticide Programs, EPA; peer reviewer of applications for Hazardous Substances Research Center Grants, National Center for Environmental Research and Quality Assurance, Office of Research and Development, EPA; peer reviewer of research conducted by the Experimental Toxicology Division, National Health and Environmental Effects Research Laboratory, EPA; peer reviewer for EPA of state-of-the-science documents including one discussing Incorporating Children's Toxicokinetic Principles into Human Health Risk Assessments; and member of an expert panel on Assessing Risks of Environmental Agents to Children, Office of Research and Development, EPA. He has served on the editorial boards of the *Journal of Toxicology and Environmental Health, Chemosphere, Toxicology, and Toxicology and Applied Pharmacology.* Bruckner's research focuses on the toxicology and toxicokinetics of solvent drugs and solvent interactions at low exposure levels and pharmacokinetic bases for susceptibility of children to insecticides and other chemicals. The relevance of experimental designs to "real life" chemical exposures is of particular interest. He has published more than 200 journal articles, book chapters, and abstracts. He has served on many National Academies Committees, including the Board on Environmental Studies and Toxicology Subcommittee on Acute Exposure Guideline Levels, the Committee on Health and Safety Consequences of Child Labor, the Committee on Pesticides in the Diets of Infants and Children, the Subcommittee on Dibromochloropane, and the Committee on Safe Drinking Water.

Alicia Carriquiry, B.S. (Ag Engineering), Universidad del Uruguay; M.Sc. (Animal Genetics), University of Illinois; M.Sc. (Statistics), Iowa State University; Ph.D. (Statistics and Animal Science), Iowa State University, is Associate Provost and Professor of Statistics, Iowa State University. She was a Visiting Professor at the Institute for Statistics and Decision Sciences, Duke University, and at the Department of Statistics, Pontifical Catholic University of Chile. She also serves as a Consultant to Mathematical Policy Research, ABT Associates, Kemin Food Industries, and Law and Economics Consulting Group. She is an Elected Member of the International Statistical Institute and a Fellow of the American Statistical Association. She is Past President of the International Society for Bayesian Analysis and serves on the Executive Committee of the Institute of Mathematical Statistics. She has been a Trustee of the National Institute of Statistical Sciences since 1997, and currently serves on its Executive Committee. She also is a member of the Board of the Plant Sciences Institute at Iowa State University. Carriquiry is Editor of Statistical Sciences and serves on the editorial boards of several Latin American journals of statistics and mathematics. She has published over 50 refereed articles and technical reports and has co-edited four books. Her research interest is in the development of Bayesian methods and on the application of those methods to problems in public health, human nutrition, genetics, and economics. She also has worked in the area of stochastic volatility and other nonlinear models for time-dependent data. She has served on two National Academies committees: the Subcommittee on Interpretation and Uses of Dietary Reference Intakes and the Committee on Evaluation of USDA's Methodology for Estimating Eligibility and Participation for the WIC Program. She has been a co-author of four National Academy of Sciences reports and is a member of the Federal Steering Committee Future Directions for the CSFII/NHANES Diet/Nutrition Survey: What We Eat in America.

Ellen Wright Clayton, B.S., Duke University; M.S., Stanford University; J.D., Yale University; M.D., Harvard University, is one of the preeminent scholars in the field of law and genetics. She joined the Vanderbilt faculty in 1988 and holds appointments in both the Medical School and Law School. She is the Director of the Genetics and Health Policy Center at Vanderbilt and holds the Rosalind E. Franklin Chair in Genetics and Health Policy there. She has published two books and has authored numerous chapters and articles in medical journals, interdisciplinary journals, and law journals on the intersection of law, medicine, and public health. Clayton has collaborated with faculty in the Law School, Medical School, and Sociology Department in producing interdisciplinary research. She has been an active participant in policy debates advising the National Human Genome Research Institute as well as numerous national and international bodies concerned with the ethical conduct of research involving humans for many years. She is currently the Co-chair of the Ethical, Legal, and Social Issues Committee of the International HapMap Project as well as its liaison to Japan. In addition to teaching at the Law School and Medical School, Clayton is a practicing pediatrician at the Vanderbilt Medical Center. She currently serves as a member of the Institute of Medicine's Board on Health Sciences Policy.

John Doull, B.S. (Chemistry), Montana State University; Ph.D. (Pharmacology), University of Chicago; M.D., University of Chicago, is Professor Emeritus of Pharmacology and Toxicology and Therapeutics, Department of Pharmacology, Toxicology and Therapeutics, University of Kansas Medical Center. Prior to that, he was Assistant Director of the University of Chicago Toxicity Laboratory and Associate Professor in the Department of Pharmacology at the University of Chicago. He served on the Toxicology Study Section of the National Institutes of Health and the Council of the National Institute of Environmental Health Sciences. He is past president of the Society of Toxicology and the American Board of Toxicology, has chaired the Threshold Limit Value Committee of the American Conference of Governmental Industrial Hygienists, served on the Expert Panels of the International Life Sciences Institute, the Federal

Emergency Management Agency, and DISCUS, and was a member of the Presidential Clean Air Commission. He has served on the scientific advisory panels of EPA, the National Institute for Occupational Safety and Health, and others, and he consults with many governmental, state, industrial, and private organizations. He has received numerous awards from the Society of Toxicology, Robert Wood Johnson Medical School, International Society for Regulatory Toxicology and Pharmacology, Department of the Army, University of Chicago, American Conference of Governmental Industrial Hygienists, and American College of Toxicology. Doull currently serves on the National Academies Board on Environmental Studies and Toxicology and the Subcommittee on Acute Exposure Guidelines Levels. He has also served on the Committee on Risk Assessment of Exposure to Radon in Drinking Water (Chair), the Committee on Interactions of Drugs, Biologics, and Chemicals in U.S. Military Forces, the Committee on Risk Assessment of Hazardous Air Pollutants, the Committee on Risk Assessment Methodology, the Subcommittee on Guidelines for Developing Community Emergency Exposure Levels for Hazardous Substances (Chair), the Committee on Toxicology (Chair), the Advisory Committee on the CDC Study of the Health of Vietnam Veterans, the Committee on Methods for In Vivo Toxicity Testing of Complex Mixtures from the Environment (Chair), the Board on Toxicology and Environmental Health Hazards, the Committee on Identification of Toxic and Potentially Toxic Chemicals for Consideration by the National Toxicology Program, and the Committee on Toxicity Data Elements.

Henry T. (Hank) Greely, A.B., Stanford University; J.D., Yale Law School, is the Director of the Stanford Center for Law and the Bioscience, the C. Wendell and Edith M. Carlsmith Professor of Law, and Professor, by courtesy, of Genetics at Stanford University. He chairs the steering committee of the Stanford University Center for Biomedical Ethics; co-directs the Stanford Program in Law, Science, and Technology; and co-directs the Stanford program on Genomics, Ethics, and Society. He specializes in legal and social issues arising from advances in the biosciences and in health law and policy and has written on issues concerning genetic testing, human cloning, the ethics of human genetics research, and policy issues in the health care financing system. Greely has been a member of the Stanford faculty since 1985 and served as Chair of the Stanford Faculty Senate (2002-2003). He serves on the California Advisory Committee on Human Cloning; the Scientific Advisory Committee and the Ethical Advisory Committee for the Veteran's Affairs Department's Program on Genetic Tissue Banking in Veteran's Affairs Clinical Research; and the North American Committee of the Human Genome Diversity Project, whose ethics subcommittee he chairs. He served as a law clerk for Judge John Minor

Wisdom on the United States Court of Appeals and for Justice Potter Stewart of the United States Supreme Court.

Siobán D. Harlow, B.A. (Health Arts and Sciences), University of California, Berkeley; Ph.D. (Epidemiology), Johns Hopkins School of Hygiene and Public Health, is Associate Professor, Epidemiology, Department of Epidemiology, School of Public Health, University of Michigan and Associate Director of the International Institute, University of Michigan. She also is Director of the Advanced Studies Center and Faculty Associate, Center for Research on Ethnicity, Culture, and Health, School of Public Health, both at the University of Michigan. In addition she is a member of the Scientific and Technical Advisory Group of the Department of Reproductive Health and Research at the World Health Organization. She was the convener of the international, interdisciplinary workshop on "Risk Assessment in the Context of Trade Disputes" and is editor of the forthcoming collection of papers to appear in *Risk Assessment: An International Journal*. She has served on numerous grant review panels for the National Institute of Environmental and Health Sciences, the National Institute for Occupational Safety and Health, the National Institute of Child Health and Human Development, and the Workplace Safety and Insurance Board of Ontario. Her research focuses on reproductive, prenatal, and occupational epidemiology in developing countries. She has helped develop a generation of reproductive epidemiologists in Mexico who focus on the adverse effects of environmental and occupational exposures. In collaboration with El Colegio de Sonora, she co-founded the Programa de Formación de Investigadores en Salud Reproductiva to foster the development of human resources in reproductive health research in the U.S.-border region of Mexico with support from the Fogarty International Center. In collaboration with her Mexican colleagues, she has conducted some of the first epidemiologic studies of the health status of the maquiladora workers, evaluating the interlinkages between export-led development strategies and health. Her memberships include Phi Beta Kappa, Delta Omega, the North American Menopause Society, and the Society for Epidemiologic Research.

Lester B. Lave (IOM), B.A. (Economics), Reed College; Ph.D. (Economics) Harvard University, is the Harry B. and James H. Higgins University Professor of Economics and Finance; Professor, Engineering and Public Policy, the H. John Heinz III School of Public Policy and Management; Director, Green Design Initiative; and Co-director, Carnegie Mellon University Electricity Industry Center, Carnegie Mellon University. His work has focused on environmental quality, risk perception and communication, and risk analysis and risk management—devising tools that quantify health, safety, and environmental risks and then investigating ways to manage these risks more efficiently and effectively. He has examined the effects of air pollution on human health and developed air pollution policy that is both efficient and effective and evaluated the information content of tests for determining whether chemicals are toxic and the value of tests in reproductive toxicology. He is the recipient of the Distinguished Achievement Award of the Society for Risk Analysis. Lave has served on committees of the American Medical Association and the American Academy for the Advancement of Science, participating as Acting Chairman of the Assembly of Social and Behavioral Sciences. He has participated on many grant review panels of the National Institutes of Health, the National Science Foundation, and EPA. He has served on numerous Academy committees, including the Committee on Risk-Based Analysis for Flood Damage Reduction, the Committee on Industrial Competitiveness and Environmental Protection, the Committee on the Medical Use Program of the Nuclear Regulatory Commission, the Board on Natural Disasters, the Board on Health Promotion and Disease Prevention, the U.S. National Committee for the Decade for Natural Disaster Reduction, the Committee on Dietary Guidelines Implementation, the Water Science and Technology Board, the Committee on Dam Safety, and the Energy Engineering Board.

Bernard Lo (IOM), A.B. (Physics), Harvard College; M.A. (Comparative Literature), University of Sussex; A.M. (History of Science), Harvard University; M.D., Stanford University, is Professor of Medicine and Director, Program in Medical Ethics, University of California, San Francisco. He directs the Greenwall Faculty Scholars in Bioethics Program and is a member of the Recombinant DNA Advisory Committee at the National Institutes of Health (NIH) and the Data Safety Monitoring Board for NIH-sponsored clinical trials in diabetes. Lo formerly was a member of the National Bioethics Advisory Commission and the Data Safety Monitoring Board for the AIDS Clinical Trials Group at NIH. He also directed the national coordinating office for the Robert Wood Johnson Foundation Initiative to Strengthen the Patient-Provider Relationship in a Changing Health Care Environment, and he chaired the End-of-Life Committee convened by the American College of Physicians. He is a former member of the Board of Directors of the American Society of Law, Medicine, and Ethics and the American Society for Bioethics and Humanities. He has written more than 100 articles in peer-reviewed medical journals on issues such as decisions about life-sustaining interventions, decision making for incompetent patients, physician-assisted suicide, ethical issues regarding HIV infection, and the doctor-patient relationship in managed care. He is the author of *Resolving Ethical Dilemmas: A Guide for Clinicians,* a comprehensive analysis of ethical dilemmas in adult medicine. He also is a practicing general internist who teaches clinical medicine to residents and medical students. He is a member of the Institute of Medicine (IOM) and serves on the IOM Council and on the Report Review Committee of the National Research Council. He formerly was a member of the IOM Board on Health Sciences Policy, which he chaired from 1999 to 2002. He also chaired the Committee on the Role of Institutional Review Boards in Health Services Research Data Privacy Protection.

Thomas A. Louis, B.A. (honors in Mathematics), Dartmouth College; Ph.D. (Mathematical Statistics), Columbia University, is Professor, Department of Biostatistics, Johns Hopkins Bloomberg School of Public Health. He was Assistant Professor, Department of Mathematics, Boston University (1973-1978); Associate Professor, Department of Biostatistics, Harvard School of Public Health (1978-1987); Professor, Division of Biostatistics, University of Minnesota School of Public Health (1987-2000, Division Head 1987-1999); Senior Statistical Scientist, RAND (2000-2002), and Visiting Scholar, Committee on National Statistics (CNSTAT), National Academy of Sciences (1999). He is an elected member of the International Statistical Institute and a Fellow of the American Statistical Association and of the American Association for the Advancement of Science. He served as President of the Eastern North American Region of the International Biometrics Society and is Chair-elect of the American Statistical Association's Section on Bayesian Statistical Science. From 2001 through 2003, he was coordinating editor of the *Journal of the American Statistical Association*. He serves on the Health Review Committee of the Health Effects Institute. Louis has published more than 150 articles, books, and monographs. His research interests include risk assessment; environmental, health, and public policy and the development of related statistical approaches. He concentrates on Bayesian modeling, including small area estimation, the analysis of observational studies, and research synthesis. Current applications include assessing the health effects of airborne particulate matter, assessing the cardio-pulmonary consequences of AIDS therapies, and reproductive health and the evaluation of teacher effectiveness. His Academy service includes membership on CNSTAT and on the Board of the Institute of Medicine's (IOM's) Medical Follow-up Agency. He served on the IOM Panel to Assess the Health Consequences of Service in the Persian Gulf War and on the CNSTAT Panel on Estimates of Poverty for Small Geographic Areas, and he chaired the CNSTAT Panel on Formula Allocation of Federal and State Program Funds.

Joseph V. Rodricks, B.S. (Chemistry), Massachusetts Institute of Technology; M.S. (Organic Chemistry), University of Maryland; Ph.D. (Biochemistry) University of Maryland, is Founding Principal, Environ International Corporation (1982). He is a Visiting Professor at the Johns Hopkins University School of Public Health. He is an internationally recognized expert in the field of toxicology and risk analysis and in their uses in regulation and in the evaluation of toxic tort and product liability cases. He has testified before Congress on risk assessment related to pesticides and food safety. Since 1980, he has consulted for hundreds of manufacturers, for government agencies and for the World Health Organization. He currently serves on Academy committees on Dietary Reference Intakes for Nutrients and Gulf War and Health. He has previously served on 16 Academy committees, including the Committee on Toxicological and Performance Aspects of Oxygenated Motor Vehicle Fuels, the Committee on Risk Assessment of Hazardous Air Pollutants, the Committee on Neurotoxicology and Models for Assessing Risk; the Committee on Human Health Risk Assessment of Using Antibiotics in Animal Feed, the Committee on Public Health Risk Assessment of Poultry Inspection, the Board on Toxicology and Environmental Health Hazards, the Subcommittee to Evaluate Effects of Short-Term Exposures to Drinking Water Contaminants (Chair), and the Committee on Institutional Means for Assessment of Risks to Public Health. He has written more than 100 publications on toxicology and risk analysis and has lectured nationally and internationally on these topics. Recent articles and book chapters include "Some Attributes of Risk Influencing Decision-Making by Public Health and Regulatory Officials" and "Toxicological Risk Assessment in the Courtroom: Are Available Methodologies Suitable for Evaluating Toxic Tort and Product Liability Claims?" Rodricks was formerly Deputy Associate Commissioner, Health Affairs, and Toxicologist, Food and Drug Administration. He is a Diplomat, American Board of Toxicology. His experience includes chemical products and contaminants in foods, food ingredients, air, water, hazardous wastes, the workplace, consumer products, and medical devices and pharmaceutical products. He is the author of *Calculated Risks*, a nontechnical introduction to toxicology and risk analysis, now in its sixth printing, for which he won an award from the American Medical Writers Association.

Christopher H. Schroeder, B.A., Princeton University; M.Div., Yale University; J.D., University of California, Berkeley, is Charles S. Murphy Professor of Law and Public Policy Studies and Director of the Program in Public Law, Duke University Law School. He has served as Acting Assistant Attorney General in the Office of Legal Counsel at the Department of Justice. He also has served as Chief Counsel to the Senate Judiciary Committee. His areas of research and scholarship include environmental and administrative law, democratic theory, legislative institutions, and separation of powers. He has taught environmental law; government, business and public policy; environmental litigation; toxic substances regulation; and philosophy of environmental protection. He has written on the philosophical foundations of risk regulation and liability, the regulation of toxic substances, the performance of American environmental policy, and a variety of topics in public law and theory. He co-authored a leading environmental law casebook, *Environmental Regulation: Law, Science, and Public Policy.* He is the editor of forthcoming Resources for the Future book evaluating the performance of EPA. He has written extensively on environmental and administrative law, risk regulation and liability, and regulation of toxic substances.

Robert Temple, B.A., Magna Cum Laude, Harvard College; M.D., New York University School of Medicine. At New York University, he was elected to Alpha Omega Alpha, and he completed an internship and residency in internal medicine at the Columbia Presbyterian Medical Center in 1969. Board certified in internal medicine and clinical pharmacology, Temple is Director of the Office of Medical Policy of the Food and Drug Administration's (FDA's) Center for Drug Evaluation and Research and Acting Director of the Office of Drug Evaluation 1 (ODE-1). ODE-1 is responsible for the regulation of cardio-renal, oncologic, and neuropharmacologic/psychopharmacologic drug products. The Office of Medical Policy is responsible for regulation of promotion through the Division of Drug Marketing, Advertising, and Communication and for assessing quality of clinical trials and helping to assure participant protection through the Division of Scientific Investigations. Temple has a longstanding interest in the design and conduct of clinical trials and has written extensively on this subject, especially on choice of control group in clinical trials, evaluation of active control trials, trials to evaluate dose-response, and trials using "enrichment" designs. He was Clinical Associate and then Chief Clinical Associate in the Clinical Endocrinology Branch of the National Institute of Arthritis, Metabolism, and Digestive Diseases, National Institutes of Health, from 1969 to 1972, investigating the effects of lithium on the thyroid and examining the effects of agents that disrupt microtubules on steroid secretion. He became a reviewing Medical Officer in the Division of Metabolic and Endocrine Drug Products in 1972 and became Assistant to the Director of the Bureau of Drugs in 1974. In 1976, he became the Director of the Division of Cardio-Renal Drug Products, serving in that role until 1982. From 1982 to 1988 he was Acting Director and then Director of the Office of Drug Research and Review. Among other awards, he has received FDA's Award of Merit on six occasions, three Commissioner's Special Citations, the Public Health Service Superior Service award, the Department of Health and Human Services

Distinguished Service Award, the Secretary's Special Citation, and the Drug Information Association Outstanding Service Award. He received the American Society for Clinical Pharmacology and Therapeutics' Rawls-Palmer Progress in Medicine Lecture and Award in 2001. He also received the National Organization for Rare Disorders Public Health Leadership Award in 2001. In 2002, he received the Food and Drug Law Institute's Distinguished Service and Leadership Award. He is on the editorial board of the journal *Clinical Pharmacology and Therapeutics.* He was on the Board of Directors of the Society for Clinical Trials from 1983 to 1987 and was President of the Society in 1987. He is an honorary Fellow of the American College of Clinical Pharmacology.

LIAISON

David Korn (IOM), B.A. Harvard College; M.D., Harvard Medical School, is Senior Vice President for Biomedical and Health Sciences Research, Association of American Medical Colleges. He is a former Carl and Elizabeth Naumann Professor, Dean of Stanford University's School of Medicine, and Vice President of the University, as well as former Department of Pathology Professor and Chairman and Physician-in-Chief, Pathology, at Stanford University Hospital. He has served as the chair of the National Cancer Advisory Board (presidential appointment) and also the Food and Drug Administration's Science Board's Subcommittee to Review the Intramural Research Program. He was a member of the President's Committee of Advisers on Science and Technology's Panel on Health Care Reform and serves on the Department of Veterans Affairs National Research Advisory Council. He is a fellow and member of the American Association for the Advancement of Science Council, past President of the American Society of Investigative Pathology, former President of the Association of Pathology Chairmen, honorary Fellow of the American Society of Clinical Pathologists, and Fellow of the College of American Pathologists. He has held editorial positions on *Human Pathology, American Journal of Pathology,* and the *Journal of Biological Chemistry.* He is a member of the Institute of Medicine and the National Academies Science, Technology, and Law Panel.

STAFF

Anne-Marie Mazza, B.A., Economics; M.A., History and Public Policy; Ph.D., Public Policy, the George Washington University, joined the National Academies in 1995 and has served as Senior Program Officer with both the Committee on Science, Engineering, and Public Policy and the Government-University-Industry Research Roundtable. In 1999 she was named the first director of the Science, Technology, and Law Program. Between October 1999 and October 2000, she divided her time between the Science, Technology, and Law Program and the White House Office of Science and Technology Policy, where she served as a Senior Policy Analyst.

Michelle C. Catlin, M.Sc., Pharmacology and Toxicology, Queen's University, Canada; Ph.D., Environmental Health-Toxicology Program, University of Washington, also is Senior Program Officer for the Institute of Medicine (IOM) Board on Health Promotion and Disease Prevention. Before joining IOM, she served as a Program Officer with the Board on Environmental Studies and Toxicology of the National Research Council. She has worked on numerous National Academies reports, including *Copper in Drinking Water, Toxicological Effects of Methylmercury, Arsenic in Drinking Water: 2001 Update*, and *Veterans and Agent Orange: Update 2000* and *Update 2002*.

Kathi E. Hanna, M.S., Ph.D., is a science and health policy consultant, writer, and editor in the Washington, D.C., area specializing in biomedical research policy and bioethics. She has served in senior staff and consulting positions with the National Bioethics Advisory Commission, the Presidential Advisory Committee on Gulf War Veterans Illnesses, the congressional Office of Technology Assessment, the Howard Hughes Medical Institute, the National Institutes of Health, and the Institute of Medicine.

Stacey Speer, B.S., Biomedical Engineering, University of Tennessee, joined the National Academies' Science, Technology, and Law Program in September 2002 as the Christine Mirzayan Intern. Stacey is now the Senior Program Assistant of the Science, Technology, and Law Program. She is attending the George Washington University, pursuing a Master's of Forensic Science.

Sara Davidson Maddox, M.A., is a science and health policy writer and editor, with extensive experience in the areas of bioethics, biomedical research, and health services and quality. She was editor for the National Bioethics Advisory Commission and has participated in projects for the National Institutes of Health and the Institute of Medicine.

Appendix D

Biographical Sketches of the Members of the Science, Technology, and Law Panel

Cochair, **Donald Kennedy (NAS, IOM),** A.B. (Biology), Harvard University; Ph.D. (Biology), Harvard University, is President Emeritus and Bing Professor of Environmental Science, Stanford University. He also serves as editor-in-chief of *Science*. He served as Commissioner of the Food and Drug Administration and was a member of the National Academies planning committee that initiated the 1997 Academy Symposium on Science, Technology, and Law.

Cochair, **Richard A. Merrill (IOM),** A.B., Columbia University; B.A., Oxford University; M.A., Oxford University; LL.B., Columbia University School of Law, is Daniel Caplin Professor of Law, University of Virginia Law School. From 1975 to 1977 he served as Chief Counsel to the Food and Drug Administration. He was Dean of the Law School from 1980 to 1988. Since 1991 he has been special counsel to Covington & Burling and was a member of the National Academies planning committee that initiated the 1997 Academy Symposium on Science, Technology, and Law.

Shirley S. Abrahamson, B.A., New York University; J.D., Indiana University Law School; LL.B. (American Legal History), University of Wisconsin Law School, has since 1996 served as Chief Justice, Wisconsin Supreme Court. In that capacity, she serves as the administrative leader of the Wisconsin court system. She was previously in private practice for 14 years and taught at the University of Wisconsin Law School and Marquette University Law School.

Frederick R. Anderson, Jr., B.A. (History of Science), University of North Carolina; J.D., Harvard Law School, Oxford University, is a partner of the law firm of McKenna, Long, & Aldridge, LLP in Washington, D.C. He is former Dean of the Washington College of Law at American University. He was a member of the National Academies' planning committee that initiated the 1997 Academy Symposium on Science, Technology, and Law.

Margaret A. Berger, A.B., Radcliffe College; J.D., Columbia University School of Law, is the Suzanne J. and Norman Miles Professor of Law at Brooklyn Law School in Brooklyn, New York. She has written exclusively on science and law, and in particular on three key Supreme Court cases (*Daubert, Joiner, Kumho*) dealing with evidence. She is the co-author of Weinstein's *Evidence.*

Arthur I. Bienenstock, B.S. (Physics), Polytechnic Institute of Brooklyn; M.S. (Physics), Polytechnic Institute of Brooklyn; Ph.D. (Applied Physics), Harvard University, is Vice Provost and Dean of Research and Graduate Policy, Stanford University. He is immediate past Director of Geballe Laboratory for Advanced Materials, Stanford University. Previously he was Associate Director for Science, Office of Science and Technology Policy, Executive Office of the President (1997-2000); Director of the Stanford Synchrotron Radiation Laboratory, Stanford University (1978-1997); Vice Provost for Faculty Affairs, Stanford University (1972-1977); member of the Stanford University faculty since 1967.

Paul D. Carrington, B.A., University of Texas; LL.B., Harvard University, is Professor of Law, Duke University Law School. He is the former Dean of Duke University Law School and has taught and published extensively on civil procedures. He was Reporter to the Advisory Committee on Civil Rules of the Judicial Conference of the United States. He also established the Private Adjudication Center, which developed a Registry of Independent Scientific and Technical Advisors to provide disinterested advice to lawyers and judges on scientific issues that are the subject of legal disputes.

Joe S. Cecil, Ph.D. (Psychology), Northwestern University; J.D., Northwestern University, is a Project Director in the Division of Research at the Federal Judicial Center in Washington, D.C. Currently he is directing the Center's Program on Scientific and Technical Evidence. As part of this program he is responsible for judicial education and training in the area of scientific and technical evidence and serves as principal editor of the Center's *Reference Manual on Scientific Evidence*, the primary source book on evidence for federal judges.

Joel E. Cohen (NAS), Dr.P.H. (Population Sciences and Public Health), Harvard University; Ph.D. (Applied Mathematics), Harvard University, is the Abby Rockefeller Mauzé Professor of Populations at the Rockefeller University. At Columbia University, he is Professor of Populations in the School of International and Public Affairs, the Department of Earth and Environmental Sciences, and the Department of Ecology, Evolution, and Environmental Biology. He heads the Laboratory of Populations at both Rockefeller and Columbia. From 1991 to 1995 he served as a U.S. Federal Court-appointed neutral expert on projections of asbestos-related claims associated with the Manville Personal Injury Settlement Trust. In addition, he has served as a Special Master in silicone gel breast implant products liability.

Kenneth W. Dam, B.S., University of Kansas; J.D., University of Chicago; LL.D. (honorary), New School for Social Research, is Max Pam Professor of American and Foreign Law, University of Chicago Law School, and a Senior Fellow at the Brookings Institution. He has devoted his career to public policy issues, both as a practitioner and as a professor. In the former capacity he served as Deputy Secretary, Department of the Treasury (2001-2003) and in the Department of State (1982-1985). In 1973 he was Executive Director of the Council on Economic Policy, a White House office responsible for coordinating domestic and international economic policy. Most of his academic work has centered on law and economics, particularly with respect to international issues.

Rebecca S. Eisenberg, J.D., is Robert and Barbara Luciano Professor of Law at the University of Michigan in Ann Arbor, Michigan. She regularly teaches courses in intellectual property, patent law, trademark law, and torts and has taught courses on legal regulation of science and on legal issues associated with the Human Genome Project.

David J. Galas, A.B. (Physics), University of California-Berkeley; M.S. (Physics), University of California, Davis-Livermore; Ph.D. (Physics), University of California, Davis-Livermore, is Chancellor, Chief Scientific Officer, and Norris Professor of Applied Life Sciences, Keck Graduate Institute of Applied Life Sciences, Claremont, California. He has a unique mix of experience in business, government, and the academic world and has most recently served as President and Chief Scientific Officer of Seattle-based Chiroscience R & D Inc., a company with an integrated approach to drug discovery. Chiroscience R & D Inc., was formed through the acquisition of Darwin Molecular Corporation, which Galas helped start.

David L. Goodstein, Ph.D. (Physics), University of Washington, is Vice

Provost and Professor of Physics and Applied Physics of the California Institute of Technology, where he has been on the faculty for more than 35 years. His book, *States of Matter*, helped launch a new discipline, condensed matter physics. He has turned his attention to societal issues that affect science as a profession. In articles, speeches, and colloquia he has addressed conduct and misconduct in science, the end of exponential growth of the scientific enterprise, and issues related to fossil fuel and the climate of Planet Earth.

Sheila S. Jasanoff, Ph.D., Harvard University; J.D., Harvard University, is Pforzheimer Professor of Science and Technology Studies at Harvard University's John F. Kennedy School of Government. Her longstanding research interests center on the interactions of law, science, and politics in democratic societies. She has written extensively on the place of science and technology in U.S., European, and Indian politics, including *Controlling Chemicals: The Politics of Regulation in Europe and the U.S.; Risk Management and Political Culture; The Fifth Branch: Science Advisers as Policymakers;* and *Science at the Bar: Law, Science, and Technology in America.*

Daniel J. Kevles, B.A. (Physics), Princeton University; (European History) Oxford University; Ph.D. (History), Princeton University, is Stanley Woodward Professor of History, Yale University, and J.O. and Juliette Koepfli Professor of the Humanities Emeritus at the California Institute of Technology. His research interests and extensive writing include the interplay of science and society past and present; history of science in America; history of modern physics; history of modern biology; scientific fraud and misconduct; and the history of intellectual property in living organisms.

David Korn (IOM), B.A., summa cum laude, M.D., cum laude, Harvard University, is Senior Vice President for Biomedical and Health Sciences Research at the Association of American Medical Colleges in Washington, D.C. Previously, he served as Carl and Elizabeth Naumann Professor and Dean of the Stanford University School of Medicine. In recent years he has written and spoken about issues of health and science policy, topics in which he has been heavily engaged on the national scene.

Robert A. Lonergan, A.B. (English Literature), Fordham College; J.D., Fordham University School of Law (1975); Finance for Senior Executives, Harvard Business School (1997), is Vice President, General Counsel and Corporate Secretary, Rohm and Haas. He is responsible for all of the company's legal affairs, directing the efforts of 45 Rohm and Haas in-house attorneys, and scores of outside law firms. He is responsible for ensuring that the company continues to meet compliance, regulatory, safety, environmental, and employment law obligations in the more than 100 countries in which Rohm and Haas conducts business.

Patrick A. Malone, J.D., Yale Law School, is a partner in the law firm of Stein, Mitchell & Mezines in Washington, D.C. Malone, a former medical journalist, represents plaintiffs in medical malpractice and product liability lawsuits. He is a member of the Association of Trial Lawyers, Trial Lawyers of America, and Trial Lawyers for Public Justice.

Richard A. Meserve (NAE), J.D., Harvard Law School; Ph.D. (Applied Physics), Stanford University, is President, Carnegie Institution of Washington. Before assuming the Carnegie presidency in April 2003, he was Chairman of the Nuclear Regulatory Commission, having served since October 1999. He was a member of the National Academies planning committee that initiated the 1997 Academy Symposium on Science, Technology, and Law, and he wrote the *amicus* briefs on behalf of the National Academy of Engineering in the *Kumho* case and on behalf of the National Academy of Sciences in the *Daubert* case. These landmark cases established the basis for admitting expert testimony into court.

Alan B. Morrison, LL.B., Harvard Law School, is with Public Citizen Litigation Group in Washington, D.C., a nonprofit citizen research, lobbying, and litigation organization. Prior to his work at Public Citizen, he was an associate in a law firm and an Assistant U.S. Attorney in New York City.

Thomas D. Pollard (NAS, IOM), M.D., Harvard Medical School, is the Eugene Higgins Professor, Department of Molecular, Cellular and Developmental Biology, Yale University. He was the first to elucidate the diversity of myosin motor proteins and is an expert in the biochemistry and cell biology of proteins that control the dynamics of the actin cytoskeleton.

Channing R. Robertson, B.S. (Chemical Engineering), University of California, Berkeley; M.S. (Chemical Engineering), Stanford University; Ph.D. (Chemical Engineering—emphasis on fluid mechanics and transport phenomena), Stanford University, is the Ruth G. and William K. Bowes Professor and also Dean of Faculty and Academic Affairs, School of Engineering and Professor, Department of Chemical Engineering, Stanford University. He is Director of the Stanford University-National Institutes of Health Graduate Training Program in Biotechnology. Because of his interest in biotechnology, he has consulted widely in the design of biomedical diagnostic devices.

Jonathan M. Samet (IOM), A.B. (Chemistry and Physics), Harvard College; M.S. (Epidemiology), Harvard School of Public Health; M.D. (Medicine), University of Rochester School of Medicine and Dentistry, is Professor and Chairman, Department of Epidemiology, Johns Hopkins Bloomberg School of Public Health since 1994 and is Co-director, Risk Sciences and Public Policy Institute. An epidemiologist and pulmonary physician, he has focused on the effects of inhaled pollutants, respiratory diseases, cancer, and risk assessment. He has worked extensively on risks posed by indoor and outdoor air pollution.

Fern M. Smith, B.A. (with distinction), Stanford University, J.D., Stanford Law School, is U.S. District Judge for the U.S. District Court for the Northern District of California. Judge Smith is the author of two chapters in the third edition of *Moore's Federal Practice* and has written and spoken extensively on evidentiary matters, trial practice, and other topics.

James Gustave Speth, B.A., Yale University; M. Litt., Oxford University; J.D., Yale University, is Dean and Professor in the Practice of Environmental Policy and Sustainable Development. He served most recently as Administrator of the United Nations Development Program and Chair of the United Nations Development Group. Throughout his career, he has provided leadership and entrepreneurial initiatives to many task forces and committees whose roles have been to combat environmental degradation, including the President's Task Force on Global Resources and Environment; the Western Hemisphere Dialogue on Environment and Development; and the National Commission on the Environment.

David S. Tatel, B.A., University of Michigan; J.D., University of Chicago Law School, is Judge, U.S. Court of Appeals for the District of Columbia Circuit. He was nominated by President Clinton and commissioned in 1994. Prior to his appointment he was a partner in the Washington Law firm of Hogan & Hartson, where he managed the firm's Education Group. As head of this group he provided legal counsel to school districts, colleges, universities, and education associations throughout the country.

Sheila E. Widnall (NAE), B.S. (Aeronautics and Astronautics); M.S. (Aeronautics and Astronautics); and D.Sc., Massachusetts Institute of Technology (MIT), is Institute Professor, MIT. She previously served as Secretary of the Air Force from 1993 to 1997 and had served on the U.S. Air Force Academy Board of Visitors. A professor of aeronautics and astronautics, she is internationally known for her work in fluid dynamics.

STAFF

Anne-Marie Mazza, Director
Stacey Speer, Senior Project Assistant

Appendix E

Meeting Agendas

Meeting 1
December 16-17, 2002

8:20 *Welcome and Introductions*
Jim Childress and Mike Taylor
Committee Cochairs

8:30 *Charge to the Committee*
Stephen L. Johnson
Assistant Administrator
Office of Prevention, Pesticides and Toxic Substances
U.S. Environmental Protection Agency

9:30-1:15 *Closed Session*

1:15 *Policies, Protocols, Guidelines Governing Research with Human Participants*
Nancy E. Kass
The Phoebe R. Berman Professor of Bioethics and Public Health
Professor, The Bioethics Institute
Director, Program in Law, Ethics and Health
Johns Hopkins University Bloomberg School of Public Health

2:15 *Joint SAB/SAP Report: Comments on the Use of Data from the Testing of Human Subjects*
Christopher J. Portier
Director
Environmental Toxicology Program
National Institute of Environmental Health Sciences

3:45 *Break*

4:00 *Closed Session*

5:30 *Adjourn*

Tuesday, December 17, 2002

8:30 *Overview of FIFRA/FQPA*
Stanley H. Abramson
Partner
Arent, Fox, Kintner, Plotkin & Kahn, PLLC

Erik D. Olson
Senior Attorney
Natural Resources Defense Council

10:30 *Human Subjects Research in Environmental Policy*
Penny Fenner-Crisp
Executive Director
ILSI Risk Science Institute

11:15 *Break*

11:30 *Using Human Data in the Assessment and Management of Risk*
Michael D. Dourson
Director
Toxicology Excellence for Risk Assessment

12:15 *Lunch*

1:00 *Pesticides in the Diets of Infants and Children*
Philip J. Landrigan
Ethel H. Wise Professor of Community Medicine
Chairman, Department of Community and Preventive Medicine
Director, Center for Children's Health and the Environment
Mt. Sinai School of Medicine

1:40 *Small Clinical Trials: Issues and Challenges*
Suzanne T. Ildstad
Director, Institute for Cellular Therapeutics
Professor of Surgery, Jewish Hospital
Distinguished Professor of Transplantation
University of Louisville

2:20 *Preserving Public Trust: Accreditation and Human Research Participant Programs and Responsible Research: A Systems Approach to Protecting Research Participants*
Daniel Federman
Senior Dean for Alumni Relations and Clinical Teaching
Carol W. Walter Distinguished Professor of Medicine and Medical Education
Harvard Medical School

3:00 *Break*

3:15 *Joint SAB/SAP Report: Minority Statement*
Herbert L. Needleman
Professor of Pediatrics
Member, Research Group on Lead
University of Pittsburgh Medical School

4:00-5:45 *Closed Session*

PUBLIC FORUM
JANUARY 8, 2003

8:30 *Welcome and Purpose of Forum*
Jim Childress and Mike Taylor
Committee Cochairs

8:45 *Public Input Session*
Moderator
Mike Taylor
Cochair

Speakers
8:50 Gail Charnley
HealthRisk Strategies

Jacqueline Patterson
Toxicology Excellence for Risk Assessment

9:10 Richard Wiles
Vice President for Research
Environmental Working Group

9:30 William Kelly
Western Representative
The Center for Regulatory Effectiveness

9:50 Jennifer Sass
Senior Scientist
Natural Resources Defense Council

10:10 *Break*

10:20 Ray McAllister
Vice President, Science and Regulatory Affairs
CropLife Association

10:40 Lynn Goldman
Chairman of the Board
Children's Environmental Health Network
Professor, Environmental Health Sciences
Johns Hopkins Bloomberg School of Public Health

11:00 Shelley Davis
Co-Executive Director
Farm Worker Justice Fund

11:20 Alan Herbert Lockwood
Physicians for Social Responsibility
Professor of Neurology, Nuclear Medicine, and Communicative Disorders and Sciences
State University of New York, Buffalo, School of Medicine and Biomedical Sciences

11:40 Vera Hassner Sharav
President
Alliance for Human Research Protection

12:00 *Lunch*

12:50 *Five Minute Remarks from Registered Attendees Requesting An Opportunity to Speak Before the Committee* *Moderator*
Jim Childress
Cochair

Speakers
Judith A. MacGregor
Toxicology Consulting Service

Steven H. Lamm
Consultants in Epidemiology and Occupational Health, Inc.

1:50 *Break*

2:00 *Examination of the Use of Human Toxicity Studies by Industry*
Moderator
Jim Childress
Cochair

The Role of Human Volunteer Studies in a Tiered Approach to Safety Assessment
Monty Eberhart
Director, Product Safety Management
Bayer CropScience

2:30 *Design and Conduct of Human Volunteer Studies: Ethics and Methodologies*
Angus Cameron
BCG Europe
(representing Inveresk Research International)

3:00 *Size and Statistical Power in Human Safety Studies*
Bob Sielken
Sielken and Associates Consulting, Inc.

3:15 *Case Studies*

Aldicarb
Neil Carmichael
Global Director of Toxicology
Bayer CropScience

Malathion
Chris F. Wilkinson
Principle
C. Wilkinson, LLC

Dichlorvos
Ian Chart
Vice President, Director of Regulatory Affairs
AMVAC Chemical Corporation

Perchlorate
Steven H. Lamm
Consultants in Epidemiology and Occupational Health, Inc.

Summary Remarks
Abraham J. Tobia
Regulatory Toxicologist—NFTA
Toxicology Group
Bayer CropScience

4:30 *Questions/Comments*

5:00 *Adjourn*

MEETING 2
JANUARY 9, 2003

8:30 *Welcome*
Jim Childress and Mike Taylor
Committee Cochairs

8:35 *An Overview of Risk Assessment at EPA*
William H. Farland
Acting Deputy Assistant Administrator for Science
Office of Research and Development
U.S. Environmental Protection Agency

9:30 *Application of the Common Rule to EPA Conducted and Sponsored Research*
Peter W. Preuss
Director
National Center for Environmental Research
Office of Research and Development
U.S. Environmental Protection Agency

10:30 *Break*

10:45 *FQPA—Applying the Safety Factors*
Susan Mackris
Senior Toxicologist
Office of Pesticide Programs
U.S. Environmental Protection Agency

11:45 *Risk Assessment—Pesticide Case Study—Phosmet*
Christina B. Swartz
Senior Scientist
Office of Pesticide Programs
U.S. Environmental Protection Agency

12:45 *Lunch*

1:30 *Research with Human Research Participants—The FDA Experience*
Robert Temple
Director
Office of Medical Policy
Center for Drug Evaluation and Research

2:30-3:30 *Closed Session*

MEETING 3
MARCH 19, 2003

8:00 *Consideration, Use, and Value of Human Subjects Research to EPA Program Offices*

Rita S. Schoeny
Senior Science Advisor
Office of Water
U.S. Environmental Protection Agency

Deirdre L. Murphy
Emissions Standards Divison
Risk & Exposure Assessment Group
Office of Air Quality Planning & Standards
U.S. Environmental Protection Agency

Karen M. Martin
Group Leader, Health and Ecosystems Effects Group
Office of Air Quality Planning & Standards
U.S. Environmental Protection Agency

10:00 *Value of Human Toxicity Studies*

Ernest E. McConnell
President
ToxPath

10:45 *Break*

11:00 *Pesticides and Children: Research Challenges/Exposure Issues/ Effects of Neural Development*

John L. Adgate
Assistant Professor
School of Public Health
University of Minnesota

W. Stephen Brimijoin
Professor and Chair
Department of Molecular Pharmacology and
Mayo Clinic
Experimental Therapeutics

12:15 *Lunch*

1:30 *Ethical Issues Associated with Intentional Dosing of Human Research*

Jeffrey Kahn
Director and Professor
Center for Bioethics
University of Minnesota

Arthur Caplan
The Emanuel and Robert Hart Professor of Bioethics
Chair, Deparment of Medical Ethics
Director, Center for Bioethics
University of Pennsylvania

3:00 *Adjourn*

Appendix F

Combined Registrants List for All Meetings

Stanley Abrahamson
Partner
Arent Fox Kintner Plotkin & Kahn, PLLC

John Adgate
Assistant Professor
School of Public Health
University of Minnesota

Diane Allemang
Director, Regulatory Affairs
Cheminova

Alan Ayers
Head-State Affairs/Stewardship
Bayer CropScience

Rhoda Barnat
Managing Director
Abernathy MacGregor

Craig Barrow
Dow Chemical

Nancy Beck
Toxicologist/Risk Assessor
Office of Information and Regulatory Affairs
Office of Management and Budget

Sharon Begley
Reporter
Wall Street Journal

Richard B. Belzer
President
Regulatory Checkbook

Karin Bentley
Global Regulatory Toxicologist
DuPont Crop Protection

Richard Bissell
Executive Director
Policy and Global Affairs Division
The National Academies

Ann Bleacker
Head, Regulatory Toxicology
Bayer CropScience

Elizabeth Boa
Senior Manager
Regulatory/Technical Center for Expertise
American Chemistry Council

Daniel Botts
Director
Florida Fruit and Vegetable Association
Environmental and Pest Management Division

W. Stephen Brimijoin
Professor and Chair
Department of Molecular Pharmacology and Experimental Therapeutics
Mayo Clinic

Angus Cameron
BCG-Europe
Roslin BioCentre

Lisa Campbell
Bergeson & Campbell, P.C.

Arthur Caplan
Chair, Department of Medical Ethics
Director, Center for Bioethics
University of Pennsylvania

John Carley
U.S. Environmental Protection Agency

Wayne Carlson
Bayer CropScience

Neil Carmichael
Global Director of Toxicology
Bayer CropScience

Barry Castleman

Gail Charnley
Health Risk Strategies

Ian Chart
Vice President
Director of Regulatory Affairs
AMVAC Chemical Corporation

Ethel Chase

Peg Cherny
Bayer CropScience

Margaret Chu
Office of Research and Development
U.S. Environmental Protection Agency

David Clarke
American Chemistry Council

Greg Coffey
Bayer CropScience

James W. Conrad, Jr.
American Chemistry Council

Roger Cortesi
Senior Science Advisor
U.S. Environmental Protection Agency

Shelley Davis
Co-Executive Director
Farm Worker Justice Fund

John D. Doherty
U.S. Environmental Protection Agency

Larry Dorsey
U.S. Environmental Protection Agency

Michael L. Dourson
Director
Technology Excellence for Risk Assessment

Sidney Draggan
Senior Science and Science Policy Advisor
Office of Research and Development
U.S. Environmental Protection Agency

Angelina Duggan
Science Policy Director
Science and Regulatory
CropLife America

Monty Eberhart
Director, Product Safety Management
Product Safety Management
Bayer CropScience

Ernest Falke
U.S. Environmental Protection Agency

William Farland
Acting Deputy Assistant Administrator for Science
Office of Research and Development
U.S. Environmental Protection Agency

Daniel D. Federman
Senior Dean for Alumni Relations and Clinical Teaching
Carl W. Walter Distinguished Professor of Medicine and Medical Education
Harvard Medical School

Penny Fenner-Crisp
Executive Director
Risk Science Institute International Life Science Institute

Christina Geisert
Intern
Office of Pesticide Programs
U.S. Environmental Protection Agency

Pat Getter
CropLife America

Steven Gibb
Reporter
FDA Weekly

Lynn Goldman
Chairman of the Board, Children's Environmental Health Network
Professor, Environmental Health Sciences
Johns Hopkins Bloomberg School of Public Health

Denise Grady
Reporter
The New York Times

Linda E. Greer
Program Director
Health & the Environment
National Research Defense Council

Dawn Grodsky
Managing Editor
Clean Air Report
Inside EPA

Ephi Gur
Manager of Regulatory and Scientific Affairs
Makhteshim Agan of North America Inc.

Steven G. Gurney
Geologist
Health & the Environment Program
National Research Defense Council

Barry M. Hartman
Kirkpatrick & Lockhart, LLP

Andrew Hawkins
Reporter
Blue Sheet

John Heilprin
Reporter
Associate Press

Bette Hileman
Reporter
Chemical & Engineering News

Larry Hodges
Registration Manager
Regulatory Affairs
Bayer CropScience

David Hrdy
Biologist
Health Effect Division
Office of Pesticide Programs

Leslie J. Hushka
Scientific Associate
Business Support
ExxonMobil Biomedical Sciences, Inc.

Susan Ildstad
Director, Institute for Cellular Therapeutics
Professor of Surgery, Jewish Hospital
University of Louisville

Stephen Johnson
Assistant Administrator
Office of Prevention, Pesticides and Toxic Substances
U.S. Environmental Protection Agency

William Jordan
Senior Policy Adviser for Pesticides
Office of Pesticide Programs
U.S. Environmental Protection Agency

Jeffrey Kahn
Director and Professor
Center for Bioethics
University of Minnesota

Jocelyn Kaiser
Reporter
Science Magazine

Hannah Kamenetsky
Freelance Reporter
The Scientist

Nancy Kass
Director
Program in Law, Ethics, and Health
Johns Hopkins University Bloomberg School of Public Health

William Kelly
Western Representative
Center for Regulatory Effectiveness

Jim Kling
Reporter
Web MD

David Kramer
Editor
Science and Government Report, Technical Insights
Frost and Sullivan

Steven Lamm
Consultants in Epidemiology and Occupational Health, Inc.

Philip J. Landrigan
Ethel H. Wise Professor of Community Medicine
Mt. Sinai School of Medicine

Patrick Linehen
Abernathy MacGregor

Alan Hebert Lockwood
Professor of Neurology, Nuclear Medicine
School of Medicine and Biomedical Sciences
State University of New York, Buffalo

Joan Lowry
Reporter
Scripps Howard News Service

Judith MacGregor
Toxicology Consulting Services

Susan Makris
Toxicologist
Office of Pesticide Programs
U.S. Environmental Protection Agency

Ann Manley
Director of Toxicology
AMVAC Chemical Corporation

Karen Martin
Group Leader, Health and Ecosystems Effects Group
Office of Air Quality Planning and Standards
U.S. Environmental Protection Agency

Ray McAllister
Vice President, Science and Regulatory Affairs
CropLife America

Ernest McConnell
President
ToxPath

Elizabeth Mendez
Toxicologist
Human Health Effects Division
U.S. Environmental Protection Agency

David E. Menotti
Shaw Pittman LLP

Richard Merrill
Daniel Caplin Professor of Law
School of Law
University of Virginia

Michael Metzger
Chief, RRB1
Health Effects Division
U.S. Environmental Protection Agency

David Miller
Office of Pesticide Programs
U.S. Environmental Protection Agency

Deirdre Murphy
Emissions Standards Division
Office of Air Quality Planning & Standards
U.S. Environmental Protection Agency

Herber L. Needleman
Professor of Pediatrics
Member, Research Group on Lead
University of Pittsburgh Medical School

George Oliver
Science Policy Leader
Government and Public Affairs
Dow AgroSciences

Eric Olson
Senior Attorney
Natural Resources Defense Council

Jacqueline Patterson
Environmental Scientist
Toxicology Excellence for Risk Assessment

Pat Phibbs
Reporter
Daily Environment Report
BNA, Inc.

Mary Beth Polley
Assistant Editor
Pesticide and Toxic Chemical News
CRC Press LLC

Andrew Pope
Director, Health Science Policy Program
Institute of Medicince
The National Academies

Christopher Portier
Director
Environmental Toxicology Program
National Institute of Environmental Health Sciences

Peter Preuss
Director
National Center for Environmental Research
U.S. Environmental Protection Agency

Vivian Prunier
U.S. Environmental Protection Agency

Tony Reichhardt
Reporter
Nature

Jean Reimers
Bayer CropScience

Bruce Rodan
Medical Officer (Research)
U.S. Environmental Protection Agency

Chad Sandusky
Senior Toxicologist
Physicians Committee for Responsible Medicine

Jennifer Sass
Senior Scientist
Natural Defense Resources Defense Council

Scott Schang
Associate
Environmental, Land, and Resources Department
Latham & Watkins

Rita Schoeny
Senior Science Advisor
Office of Water
U.S. Environmental Protection Agency

Rick Schwabacher
Washington Representative
The Cousteau Society

Mark Seaton
U.S. Environmental Protection Agency

Vera Hassner Sharav
President
Alliance for Human Research Protection

Virginia Ashby Sharpe
Project Director, Integrity in Science
Center for Science in the Public Interest

Karen Shearer
Bayer CropScience
Robert L. Sielken, Jr.
Sielken & Associates Consulting, Inc.

Burleson Smith
U.S. Department of Agriculture

Carol Stroebel
Health Policy Consultant
Children's Environmental Health Network

Lauren E. Sucher
Environmental Working Group

Christina Swartz
Senior Scientist
Office of Pesticide Programs
U.S. Environmental Protection Agency

Abraham Tobia
Toxicology Fellow
Toxicology Department
Bayer CropScience

Dennis Utterback
Office of Science Policy
U.S. Environmental Protection Agency

Kenneth Weinstein
Partner
Environment, Land, and Resources Department
Latham and Watkins

Paul Whatling
Senior Product Manager
Cheminova, Inc.

Richard Wiles
Vice President for Research
Environmental Working Group

Chris F. Wilkinson
Principal
C. Wilkinson, LLC

Eric Wintemute
President
AMVAC Chemical Corporiation

Liesel Wolff
Congressional Liaison
People for the Ethical Treatment of Animals

Alison Young
Reporter
KR Washington

Susan Hunter Youngren
Bergeson & Campbell, PC